THE MINERS

THE MINERS

By the Editors of

TIME-LIFE BOOKS

with text by

Robert Wallace

TIME-LIFE BOOKS / ALEXANDRIA, VIRGINIA

Time-Life Books Inc.
is a wholly owned subsidiary of

TIME INCORPORATED

Founder: Henry R. Luce 1898-1967

Editor-in-Chief: Henry Anatole Grunwald
President: J. Richard Munro
Chairman of the Board: Ralph P. Davidson
Executive Vice President: Clifford J. Grum
Chairman, Executive Committee: James R. Shepley
Editorial Director: Ralph Graves
Group Vice President, Books: Joan D. Manley
Vice Chairman: Arthur Temple

TIME-LIFE BOOKS INC.

Managing Editor: Jerry Korn
Text Director: George Constable
Board of Editors: Dale M. Brown, George G. Daniels,
Thomas H. Flaherty Jr., Martin Mann, Philip W. Payne,
Gerry Schremp, Gerald Simons
Planning Director: Edward Brash
Art Director: Tom Suzuki
  Assistant: Arnold C. Holeywell
Director of Administration: David L. Harrison
Director of Operations: Gennaro C. Esposito
Director of Research: Carolyn L. Sackett
  Assistant: Phyllis K. Wise
Director of Photography: Dolores A. Littles

Chairman: John D. McSweeney
President: Carl G. Jaeger
Executive Vice Presidents: John Steven Maxwell,
David J. Walsh
Vice Presidents: George Artandi, Stephen L. Bair,
Peter G. Barnes, Nicholas Benton, John L. Canova,
Beatrice T. Dobie, Carol Flaumenhaft, James L. Mercer,
Herbert Sorkin, Paul R. Stewart

THE OLD WEST

EDITORIAL STAFF FOR "THE MINERS"
Editor: George G. Daniels
Picture Editor: Patricia Hunt
Text Editors: Betsy Frankel, Joan S. Reiter
Designer: Bruce Blair
Staff Writers: Don Earnest, Sally French,
Gregory Jaynes, Frank Kappler
Chief Researcher: June O. Goldberg
Researchers: Jane Coughran, Terry Drucker,
Harriet Heck, Beatrice Hsia, Jane Jordan,
Thomas Lashnits, Robert Stokes, Scot Terrell,
Gretchen Wessels
Design Assistant: Deanna Lorenz
Copy Coordinators: Barbara H. Fuller, Michele Lanning,
Susan Tribich
Picture Coordinator: Marianne Dowell

EDITORIAL OPERATIONS
Production Director: Feliciano Madrid
  Assistants: Peter A. Inchauteguiz, Karen A. Meyerson
Copy Processing: Gordon E. Buck
Quality Control Director: Robert L. Young
  Assistant: James J. Cox
  Associates: Daniel J. McSweeney, Michael G. Wight
Art Coordinator: Anne B. Landry
Copy Room Director: Susan B. Galloway
  Assistants: Celia Beattie, Ricki Tarlow

THE AUTHOR: Over the course of a long career, Robert Wallace has published more than 200 essays, poems, short stories and nonfiction articles on subjects including Admiral Hyman Rickover and the hunt for a lost Spanish treasure in Arizona. His earlier works include *The Rise of Russia* in the Great Ages of Man series, *The World of Leonardo* in the Library of Art, and in the American Wilderness series, *The Grand Canyon* and *Hawaii*. For this volume for TIME-LIFE-BOOKS, he made long research trips to Colorado, Arizona and Nevada, sites of some of the richest 19th Century gold and silver strikes.

THE COVER: Deep inside a Nevada mountain, candles affixed to shoring timbers illuminate the work of men searching for silver. This 19th Century woodcut (tinted here in the style of the period) depicts a number of arduous tasks: the men in the background slam at the walls with heavy picks while another miner cranks the windlass used to haul ore from another level. The miner in the foreground is tediously breaking up the precious metal-bearing rock. In the frontispiece, four workers, candles in hand, sit on a rail car at the mouth of a tunnel leading into the enormous Savage mine on the Comstock Lode; so deep were its shafts that the temperature in the lower levels reached a stifling 110°.

CORRESPONDENTS: Elisabeth Kraemer (Bonn); Margot Hapgood, Dorothy Bacon, Lesley Coleman (London); Susan Jonas, Lucy T. Voulgaris (New York); Maria Vincenza Aloisi, Josephine du Brusle (Paris); Ann Natanson (Rome). Valuable assistance was also provided by: Judy Aspinall, Karin B. Pearce (London); Carolyn T. Chubet, Miriam Hsia, Christina Lieberman (New York); Mimi Murphy (Rome); Janet Zich (San Francisco); Jane Estes (Seattle); Villette Harris (Washington).

Other Publications:

PLANET EARTH
COLLECTOR'S LIBRARY OF THE CIVIL WAR
LIBRARY OF HEALTH
CLASSICS OF THE OLD WEST
THE EPIC OF FLIGHT
THE GOOD COOK
THE SEAFARERS
THE ENCYCLOPEDIA OF COLLECTIBLES
THE GREAT CITIES
WORLD WAR II
HOME REPAIR AND IMPROVEMENT
THE WORLD'S WILD PLACES
THE TIME-LIFE LIBRARY OF BOATING
HUMAN BEHAVIOR
THE ART OF SEWING
THE EMERGENCE OF MAN
THE AMERICAN WILDERNESS
THE TIME-LIFE ENCYCLOPEDIA OF GARDENING
LIFE LIBRARY OF PHOTOGRAPHY
THIS FABULOUS CENTURY
FOODS OF THE WORLD
TIME-LIFE LIBRARY OF AMERICA
TIME-LIFE LIBRARY OF ART
GREAT AGES OF MAN
LIFE SCIENCE LIBRARY
THE LIFE HISTORY OF THE UNITED STATES
TIME READING PROGRAM
LIFE NATURE LIBRARY
LIFE WORLD LIBRARY
FAMILY LIBRARY:
  HOW THINGS WORK IN YOUR HOME
  THE TIME-LIFE BOOK OF THE FAMILY CAR
  THE TIME-LIFE FAMILY LEGAL GUIDE
  THE TIME-LIFE BOOK OF FAMILY FINANCE

This volume is one of a series that chronicles the history of the American West from the early 16th Century to the end of the 19th Century.

For information about any Time-Life book, please write:
Reader Information
Time-Life Books
541 North Fairbanks Court
Chicago, Illinois 60611

Library of Congress Cataloguing in Publication Data
Time-Life Books.
  The miners / by the editors of Time-Life Books; with text by Robert
  Wallace.—New York: Time-Life Books, c1976.
  240 p.: ill. (some col.); 28 cm.—(The Old West)
  Bibliography: p. 236-237.
  Includes index.
  1. Gold mines and mining—United States—History.   2. Silver mines
  and mining—United States—History.   3. Miners—United States—
  History.
  1. Wallace, Robert, 1919-   II. Title.   III. Series:
  The Old West (Alexandria, Va.)
TN413.A5T55   1976   978'.03   76-15917
ISBN 0-8094-1539-9
ISBN 0-8094-1538-0 lib. bdg.
ISBN 0-8094-1537-2 retail ed.

CONTENTS

1 | A rough way to get rich quick

"What a clover-field is to a steer, the sky to a lark, a mudhole to a hog, such are new diggings to a miner." So, in 1862, wrote *The Oregonian* about the men who roamed the West in a tireless search for precious metal. Once the great California gold strike of 1848 had shown what riches the Western earth could hold, hordes of Americans hurried to every other promising corner of the wilderness.

The grip of gold or silver fever was more than just a yearning for wealth and luxury. To many the game was the thing: a man never knew when he might spy a glint of bright metal on a remote hillside or in a virgin stream—and he never ceased expecting it to happen. A Montana gold rusher described his vice as "falling victim to *ignis fatuus,* the will-o'-the-wisp of Coeur d'Alene, Peace River, Stinkene, Cassiar, White Pine, Pioche or Yellowstone, and last but not least the Black Hills."

Through winter and summer the prospectors kept at the search; no gorge was too precarious to descend, no river too treacherous to navigate, no peak too difficult to explore. Writing from Colorado in 1860, Samuel Mallory, a gold-struck former mayor of Danbury, Connecticut, told the folks back home: "You can form no idea of the mountain we climbed (unless you can think of one ten-hundred thousand million feet high). One company of miners ran their boiler wagon off a bank 50 ft. high, and killed two pair of oxen. We feel very thankful for our safe deliverance."

A miner surveys the scene from an aerie fit for falcons at the Dolly Varden silver mine, in the Rockies near Georgetown, Colorado.

Miners gather on and around a hydraulic nozzle shooting its powerful jet of water at a gold-rich gravel bank in Boise Basin, Idaho, in the 1890s. Whole hills could be melted away by hydraulic mining, provided that quantities of water were available. This hose drew on a creek 17 miles away.

9

A burro train hauls mine-shoring timbers through Telluride, Colorado, so-named because gold in the region was combined with the element tellurium. Street lights, powered by generators of the San Miguel Gold Mining Company, gave Telluride its reputation as best-lighted town in the West.

The operators of a steam pump at the head of a shaft on Nevada's Comstock Lode are dwarfed by the machine's 40-foot flywheel and high-pressure cylinder. The pump prevented flooding in the mine by lifting as much as two million gallons of water daily 1,100 feet to a drainage tunnel.

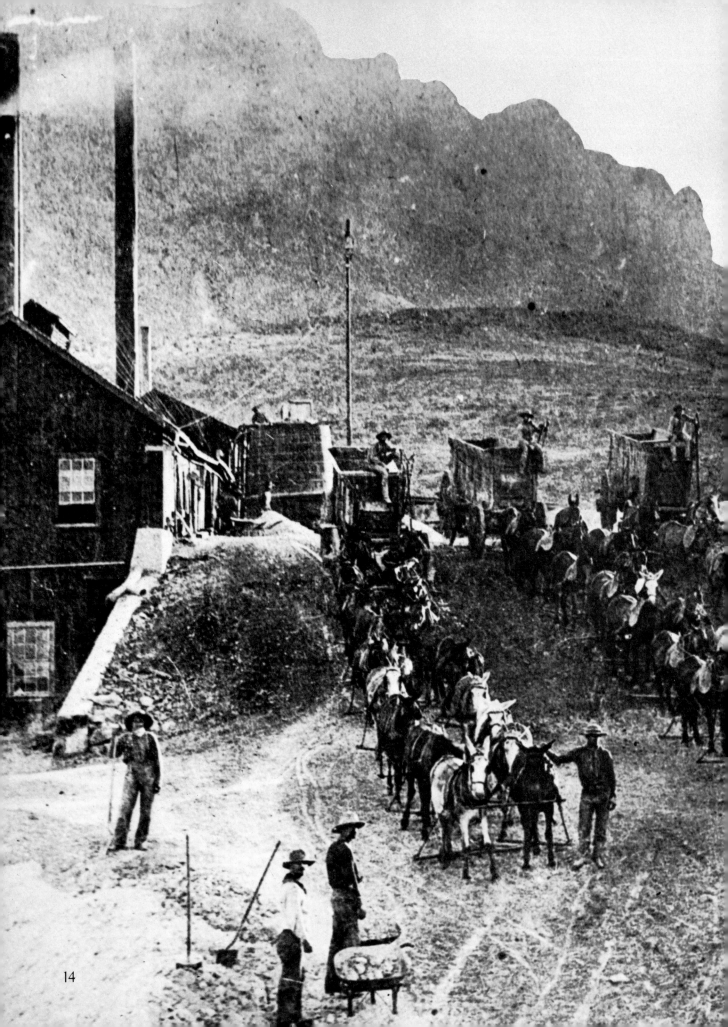

Twenty-mule-team ore wagons load up at mills of the Silver King mine in Arizona Territory. When metal-bearing rock was discovered there, claimants thought they had hit on the king of all mines; in fact, the mine yielded a treasure that was at least princely—$16 million worth of silver.

A group of miners stands outside a log building at Forty
Mile Post in Canada's Yukon Territory, while a dog team
of assorted breeds naps in the sun. Here George Carmack,
an American prospector, set off the last great gold rush in
1896 by filing a claim stating that he and two Indian companions had found gold on a branch of the Klondike River.

"Gold is found anywhere you stick your shovel"

On January 6, 1859, a lanky young prospector named George Jackson found himself dangerously short of food in the snowy Rockies 30 miles west of Denver, then a four-month-old town of 20 cabins in Kansas Territory. Jackson had just decided to quit his gold hunting and head toward Denver when he chanced upon hot mineral springs near Clear Creek. And there he saw—as he wrote in his diary—"hundreds of beautiful mountain sheep, great big bucks with curled horns, all grazing about the springs where the warm vapors had melted the snow and left the grass for them to nibble at." Jackson shot a buck and cut chops for himself and his two dogs. Then, encouraged by the hearty meal and a new food supply, he made a momentous decision: he would devote one more day to his hunt for gold.

Next morning, Jackson resumed his search along the south fork of Clear Creek, scanning the frozen stream bed for a gravel bar, where any yellow flakes from upstream might be trapped. In his travels Jackson had seen many such so-called placer (rhymes with passer) formations—all belonging to other men. He had spent a few years prospecting in California with the forty-niners, and he would recognize gold if he found it.

Late in the day Jackson spotted a promising gravel bar. He built a big bonfire on it to thaw out the surface. Because he was traveling light, without proper mining equipment, Jackson had to hack out some slushy sand with his belt knife and pan it in his tin drinking cup. He swished water around in the cup until all the light sand was washed out. Left in the cup were a few tiny but heavy yellow flakes: gold and no mistake.

Jackson eagerly panned several more handfuls of sand, collecting a vial of gold dust and one small nugget. The gold weighed about one ounce and was worth $10 at the current price in Denver.

Since he could do little work until the spring thaw, Jackson concealed the signs of his activity and marked a fir tree 76 paces to the west. Returning to camp, he wrote exultantly in his diary: "After a good supper of meat—bread and coffee all gone—I went to bed and dreamed of riches galore in that bar. If I only had a pick and pan instead of a hunting knife and the cup, I could dig out a sack full of the yellow stuff. My mind ran upon it all night long. I dreamed all sorts of things—about a fine house and good clothes, a carriage and horses, travel, what I would take to the folks down in old Missouri and everything you can think of—I had struck it rich! There were millions in it!"

George Jackson was partly right. The Clear Creek area would yield more than $100 million worth of gold in 60 years. But like most prospectors, he overestimated his personal gain. The strike did not make him rich—only a little more comfortable. On his return to the gravel bar in May, he and a few partners panned $1,900 in dust in six days' work; not long after, Jackson sold out for an unknown sum, probably modest. Nevertheless, Jackson had discovered the first major gold field in the West's immense interior wilderness, and when news of his find burst upon Denver, the results were immediate, profound and literally lifesaving.

That May, Denver and a dozen little settlements in the Rockies were inundated with confused and desperate prospectors. They had come willy-nilly in response to exaggerated reports of earlier strikes, which began appearing in Eastern newspapers in August 1858. Small amounts of gold had been found in the South Platte River—but not nearly enough to justify

Somber determination etched on his face, a gold seeker pauses for an 1859 portrait in the Pikes Peak region, site of one of the first great rushes after California and the forty-niners. The tools and massive ore sample were standard photographic props.

19

Placer miners equipped with pans, sluices and cradle-like wooden rockers spread across the prairie at gold-laced Cripple Creek, Colorado, in 1892. Many of the prospectors in the rush had been farmers—so green, went a common saying of the day, that they "mined with pitchforks."

the screaming headlines in the Kansas City *Journal of Commerce:* "THE NEW ELDORADO!!! Gold in Kansas Territory!!!"

Gold had always been a good story, and opportunists of all sorts turned it to their own advantage. In towns along the Mississippi and Missouri rivers, merchants had done a brisk business outfitting gold seekers for the 600- or 700-mile journey. No fewer than 17 writers, most of whom had never even seen the Rockies, rushed into print with guidebooks to the chimerical gold mines, which they loosely named after the region's best known terrain feature: Pikes Peak. One journalist, D. C. Oakes, extolled the ease and comfort of the trip, and another assured his readers that "Gold is found everywhere you stick your shovel."

By April 1859, a torrent of prospectors—estimated at 100,000—had set out for the "New Eldorado," most of them ill-equipped and ignorant of the hazards they faced. Their wagons, painted with the slogan "Pike's Peak or Bust," broke down in the prairie. Many "fifty-niners" got lost, or perished of thirst, hunger and disease, or got waylaid by hostile Plains Indians. About half of the emigrants never reached the Rockies, or turned back bitterly crying "fraud" after a cursory look-see failed to reveal any bonanzas. One marooned youth wrote home, "There is no gold at Pike's Peak. No man can make 10 cents a month. If you don't send me some money, I will starve to death."

By mid-May, when George Jackson was quietly beginning to exploit his find, the tide of "go-backers" reached its crest. The signs on some wagons had been changed from "Pike's Peak or Bust" to "Busted, by God." The trail to the East was marked by discarded mining tools and by dummy tombstones inscribed:

Here lies the body of D. C. Oakes,
Killed for aiding the Pike's Peak hoax.

It was at this juncture that two of Jackson's partners came down to Denver to buy more supplies. In Doyle's general store, one partner tossed a pouch full of gold dust onto the counter and said, "Here's a sample of our stuff. We're taking out nearly $2,000 a week up on the south fork of Clear Creek."

Word spread like fire in dry prairie grass. Townsmen clapped each other on the back and shouted gleefully, "Praise God, the country's saved!" And "The

21

stuff is here after all!" And "We're all right now."

Jackson's strike, and the authentic gold rush that followed it, opened up a new chapter in the saga of Western mining, and in the settlement of the American West. Now, and in the hectic half century that followed, the continent's mountain-bound interior—from the Rockies to the Sierra Nevada and the Cascades, and from Canada to Mexico—was crisscrossed by legions of prospectors and miners, who flung up hundreds of outposts in the unpopulated uplands that the forty-niners had ignored in their rush to the Pacific. "It was," wrote prospector William Parsons, "a mad, furious race for wealth, in which men lost their identity almost, and toiled and wrestled, and lived a fierce, riotous, wearing, fearfully excited life; forgetting home and kindred; abandoning old, steady habits; acquiring restlessness, craving for stimulant, unscrupulousness, hardihood, impulsive generosity, and lavish ways."

No one knew how many freelance prospectors and wage-earning miners took part in the adventure; in the race from strike to strike, foot-loose toilers of the wilderness seldom stayed put long enough to be counted. But their strikes increased at a staggering rate. By 1866, a scant seven years after Jackson's find on Clear Creek, miners had organized more than 600 far-flung mining districts in an effort to regulate their own affairs until some official government reached their remote camps. And that was merely a solid beginning. According to a careful estimate, the West may have had as many as 100,000 mining districts by 1900. Most of the strikes were small and short-lived; the boom-and-bust cycle often ran its full course in less than a decade. But in dozens of rich areas, prospectors and miners wrung enough wealth from the earth to strain their own willing credulity.

The fabulous Comstock Lode, discovered in the Washoe Mountains of Nevada in June 1859, yielded nearly $400 million in silver and gold in three decades. In Dakota's Black Hills in 1876, a prospector named Moses Manuel and his brother Fred found the Homestake mine, probably the richest single mine in the world with an output that eventually reached one billion dollars. Perhaps the densest concentration of gold ever located was discovered in 1890 at Cripple Creek, Colorado. Here, in an area six miles square, 475 mines produced $340 million in 25 years. Thanks to Cripple Creek, Clear Creek and dozens of prodigious strikes, Colorado would supplant California as the biggest gold producer in the United States before 1870.

It was because of mining—and the influx of settlers it attracted—that Colorado Territory was carved out of Kansas Territory in 1861, and became a state 15 years later. And so it went for other territories pioneered by miners. Though most of their boomtowns eventually faded into ghost towns, several camps survived to grow into cities—notably Boise, Helena and Leadville. So did Denver, Walla Walla, in Washington Territory, and other towns, which sprang to life to serve as supply depots for mining operations in the surrounding hills and high plateaus. The miners' towns were always crude and often lawless, but their historic role as advanced bases for settled societies was indispensable in many rough regions that originally had no appeal to farmers, cattlemen and merchants.

The whole phenomenon was neatly summed up in 1859 by prospector William Parsons. Writing of the first Colorado strikes, Parsons correctly prophesied, "As the discovery of gold in the mountains of California was the forerunner of an immense emigration, and the immediate cause of the erection of a new and powerful state upon the Pacific coast, so the recent discovery of the precious metal in and around the vast 'mother range' of our mountain system, is destined to exert an incalculable influence upon the growth and prosperity of the country. The Atlantic and Pacific coasts, instead of being, as they are now, divided countries, will become parts of a compact whole, joined and cemented together by bonds of mutual interest."

In 1859, conditions throughout the United States were ripe for the developments that Parsons foresaw. In the East and Middle West, the financial panic of 1857 had created a large pool of jobless workers and landless farmers—men who were free and eager to head West in search of a new life. Most of them were greenhorns, inexperienced in mining and even pioneering, and they would learn at great pain how to survive living in the wilderness.

California, meanwhile, had accumulated a large, exportable surplus of experienced prospectors and miners like George Jackson. There was plenty of gold in California; $595 million worth of it was produced by

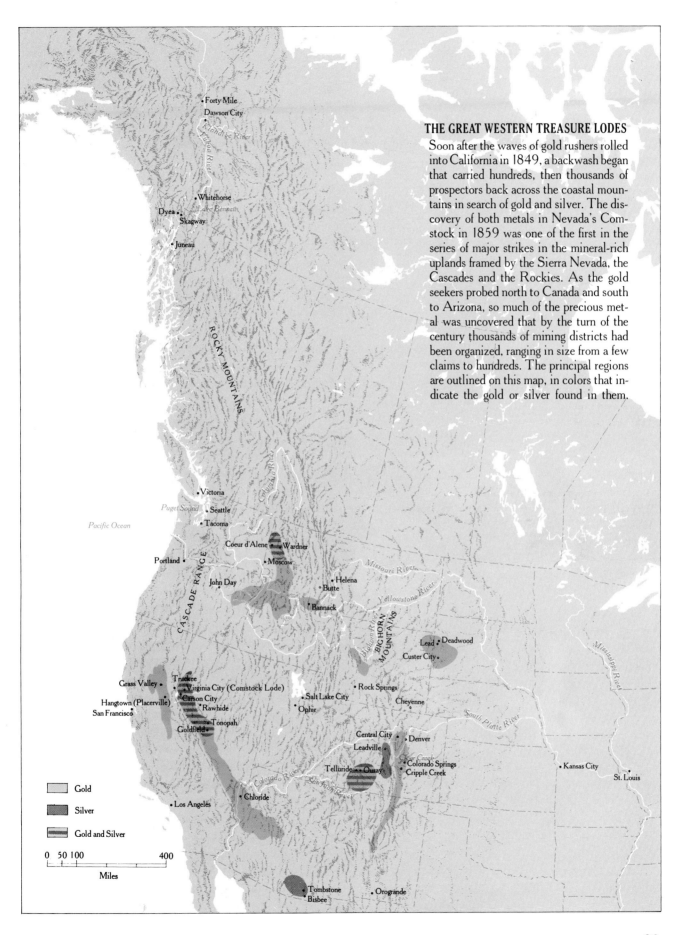

THE GREAT WESTERN TREASURE LODES
Soon after the waves of gold rushers rolled into California in 1849, a backwash began that carried hundreds, then thousands of prospectors back across the coastal mountains in search of gold and silver. The discovery of both metals in Nevada's Comstock in 1859 was one of the first in the series of major strikes in the mineral-rich uplands framed by the Sierra Nevada, the Cascades and the Rockies. As the gold seekers probed north to Canada and south to Arizona, so much of the precious metal was uncovered that by the turn of the century thousands of mining districts had been organized, ranging in size from a few claims to hundreds. The principal regions are outlined on this map, in colors that indicate the gold or silver found in them.

Forty Mile
Dawson City
Klondike River
Yukon River
Whitehorse
Lake Bennett
Dyea
Skagway
Juneau

ROCKY MOUNTAINS

Victoria
Puget Sound
Seattle
Tacoma
Pacific Ocean
Columbia River
Coeur d'Alene Wardner
Moscow
Portland
CASCADE RANGE
John Day
Helena
Butte
Bannack
Missouri River
Yellowstone River
BIGHORN MOUNTAINS
Bighorn River
Lead Deadwood
Custer City
Mississippi River

Grass Valley
Truckee
Virginia City (Comstock Lode)
Carson City
Rawhide
Hangtown (Placerville)
San Francisco
Tonopah
Goldfield
Rock Springs
Salt Lake City
Ophir
Cheyenne
South Platte River

Central City Denver
Leadville
Colorado Springs
Telluride Ouray Cripple Creek
Colorado River
San Juan River
Chloride
Kansas City
St. Louis
Los Angeles

Tombstone
Bisbee
Orogrande

Gold
Silver
Gold and Silver

0 50 100   400
Miles

23

1860, and some $700 million more would be taken out of the Sierra Nevada by 1900. But most of the placer gold, within easy reach near the surface, was quickly skimmed off. Henceforth, in the usual pattern of major strikes, mining in California was slower and more difficult, requiring massive outlays of capital to drive deep shafts into the mountains, to lay bare riverbeds by diverting streams, and to mine hydraulically, using high-pressure jets of water to break down gold-bearing ledges and even whole hills.

To make matters worse for the forty-niners, these mechanized operations employed fewer men than simple placer mining; and as the gold-rush population soared from 14,000 to 380,000 in a dozen years, more and more miners found themselves out of work or competing for low-paying, boring yet dangerous jobs deep in the bowels of the earth. The pioneer historian John S. Hittell, surveying the California gold fields in 1858, found that "The country was full of men who could no longer earn the wages to which they had become accustomed. They had become industrially desperate. They were ready to go anywhere if there was a reasonable hope of rich diggings rather than submit to live without the high pay and excitement which they had enjoyed for years in the Sacramento placers. Many of them had become unfit for the placid and orderly routine of the common laborer in other countries."

Thus it was that the Western interior was subjected to a two-pronged assault, with the seasoned miners coming from the West and supplying experience and leadership for the tenderfoot legions coming from the East. Opening new frontiers everywhere they went,

Prospectors show off pay dirt in Rockerville, South Dakota, where $350,000 worth of gold was found between 1876 and 1878.

the prospectors carried their search farther into the mountains of Colorado and Nevada and then fanned out. "Gold is where you find it," the miners often said, in mock wisdom. In the finding they poked and pried their way through the deserts of Arizona and Utah, into the wooded uplands of Washington, Oregon and Idaho, through the wilds of Montana and the Dakotas. Their explorations took them as far south as the Mexican border and as far north as the Klondike, site of the last and most arduous gold rush of the period. Yet the miners took hardships as their normal lot, and not many of the men who sampled the prospector's life willingly relinquished its freedom and its camaraderie in order to settle down.

In 1860, hard-up novices who dreamed of striking it rich were inundated with pamphlets that promised, for as little as 25 cents a copy, to tell them all they needed to know about going West and prospecting there. D. C. Oakes, the much abused and flagrantly optimistic guidebook author, was for once quite correct in advising his readers to organize in four-man "companies" for mutual protection, and to economize in buying supplies. His long list of travel essentials contained enough for six months on the trail — including three yoke of oxen, a wagon, 400 pounds of bacon, 25 pounds of gunpowder and 50 pounds of lead. Total cost: $517.25. But once a prospector had arrived on the scene, his working outfit was simpler, consisting of a pick, a shovel, a gold pan, a good rifle, food for two or three weeks and a mule or burro to carry the lot.

If a prospector lacked the cash to equip himself, he could often persuade someone in the gold camp to give

Miners use a rocker to work the San Juan's riverbed in Utah following a gold strike in 1892. Alas, most claims produced only sand.

His food gone, his oxen exhausted, his slogan an epitaph, an 1859 gold rusher expires on the plains. No one counted how many prospective miners died en route to Pikes Peak, but casualties were appalling.

him a grubstake. In frontier towns, grubstaking was a common business arrangement. A backer would supply the miners with food and tools in return for a share —often one half—of any strike the prospector made. Many merchants and saloonkeepers grubstaked likely looking prospectors. Probably the most famous of these vicarious gold hunters was H. A. W. Tabor, a cantankerous storekeeper in Leadville, Colorado, who gave a $17 grubstake to two prospectors and a year later made one million dollars on the sale of his share in the mine they found.

When prospectors ventured into new country, they looked first in the stream-bed placers, which they called poor man's mines because any gold that accumulated there could be collected without costly equipment. In fact the very name placer came from a Spanish word that in one of its meanings roughly described the feel-

ings of a placer finder—"contented" or "satisfied."

The sand and gravel placers often contained other yellowish or glittering minerals, such as iron pyrites (fool's gold) or mica, which might confuse greenhorns. But gold was unmistakable to anyone who had hefted it and clenched a bit of it between his teeth. Soft and malleable in its pure 24-carat form, gold was the only yellow metal that would not break when it was vigorously pounded or bent. Under pressure, bits of gold tended to meld into a single piece; in fact coins containing a high percentage of gold were so soft that those on the bottom of a weighty stack sometimes fused into a lump.

Under almost all conditions, pure gold stubbornly maintained its own identity. Seen from any angle, it always looked the same, whereas pyrites and mica would "wink" as light struck their various facets. Gold would not rust or tarnish, even after lying for centuries in mineral-laden water. But for all of gold's resistance to change, it did have one compliant property that aided prospectors. Gold had a strong affinity for mercury (commonly called quicksilver) and would readily amalgamate with it on contact. Thus tiny, unmanageable particles of placer gold could easily be picked up with a blob of mercury. After amalgamation, the two metals could be separated just as easily by squeezing them in a wet chamois bag; the mercury would pass out through the pores of the leather, but the gold would not.

Few prospectors had book learning in geology, but most of them absorbed enough miners' lore to understand where they were likeliest to find gold, and how it got there. They knew in a general way that, ages ago, gold-bearing rock had risen in molten form from the depths of the earth, driven upward by the violent forces that built mountains. Most of the vein matter was worthless quartz or other rock, which the miners called gangue. But enclosed in the gangue were precious metals, sometimes blended with it and sometimes occurring as separate particles and threads.

Wherever the lodes were exposed to the weather, erosion gradually broke down the gangue into crumbling chunks, then successively into fragments, sand and powder. The indestructible gold was thus released to be carried downhill by rain and mountain streams. Naturally, nuggets or large flakes of gold traveled only a short distance, while the tiny light particles called flour or flood gold went much farther — even to the ocean. The richest placers were to be found in the foothills of a mountain range, where swift streams, slowing down as they flowed into less precipitous terrain, lost their carrying power and relinquished most of their treasure. Prospectors looked particularly for locations where the rivers suddenly widened and their currents diminished sharply; for gravel bars protruding into the bend in a stream; for potholes and transverse ridges in stream-bed rock — obstacles that would serve as natural gold collectors.

Having found a promising placer, the prospector brought into play the basic tool of his trade — a gold pan. This was a three- or four-inch deep basin made of tin plate or sheet iron with sloping sides. It measured about 10 inches across its flat bottom and about 15 inches across the top. The prospector would shovel some sand into the pan, submerge the pan in the stream and spin it slowly with a flipping motion that washed the light sand and silt out over its rim. After five or 10 minutes, the pan would be washed clean of everything but a spoonful of heavy residue called the drag — probably a corruption of the word dregs. Then, with a skillful flick, the prospector fanned out the drag on the bottom of the pan, revealing, if he was in luck, a little comet tail of gold specks called colors. He picked out each color with a knife or fingernail and stashed it in a bottle or can, transferring the gleanings to a leather pouch when he quit work at nightfall.

In a long day spent squatting in a cold mountain stream, a miner could process about 50 panfuls of sand, and he would make ends meet at high boomtown prices if he averaged 10 cents worth of gold per pan. In rich placers, miners sometimes washed panfuls of sand worth $50 each, and on rare claims in the Klondike, individual pans yielded as much as $800.

If a placer's early yield was encouraging, two or more prospectors might team up to build and operate a rocker, a sluice or a long tom. These crude devices differed in shape and detail, but all worked efficiently in the same general way to help miners sift more sand faster. Essentially, each apparatus was a flat wooden trough, built on a slant, with a shallow hopper on the higher end and wood strips, or riffles, nailed across the trough. River-bank material was shoveled into the hopper, whose wire-mesh covering screened out stones and pebbles. The light sand that went through the screen was doused continuously with water and washed down and out of the trough. But the heavy gold-bearing sediment was caught behind the riffles, and the miners would scrape it out and pan it from time to time. A panful of this concentrated stuff occasionally yielded as much as $1,000 worth of gold, and veteran miners delighted in mentioning the fact casually to neophytes, who usually believed that the $1,000 came from a single swipe in a cold creek.

The gold in rich placers often warranted more complicated operations. Once gold came to rest on a river bottom, it tended to sink through lighter sand and sediment until it hit bedrock, forming what miners called a pay streak anywhere from a few inches to dozens of feet below the water course. To reach the gold, miners dammed or diverted the stream; then they would dig a shaft downward, panning samples of the sediment as they descended. If the pay streak turned out to be especially rich, they might remove all the stream-bed sediment down to bedrock.

The gold also could be collected by coyoting — digging tunnels that radiated like spokes from the bottom of a shaft. Of course, common sense dictated that any tunnel cut through soft sediment should be shored up with stout posts and caps. But most prospectors were inveterate optimists, and even with plenty of timber available, they scorned safety measures and gambled on making a quick harvest. Many a miner coyoted his way into collapsing muck that became his grave.

Though placer gold was always associated with rivers, the search for it took a number of experienced miners far from running water. They knew rivers often changed course or dried up completely, so they ventured into the desert seeking the stream beds of vanished rivers in the foothills of arid mountain ranges. Finding a placer-like ancient sand bar, but lacking wa-

Miners rest beside a sluice box in Colorado Territory's Gregory Gulch. This strike by John Gregory in 1859 revived the Pikes Peak rush.

ter to pan it for gold, the dry-country miner or desert rat would shovel sand into a blanket and toss it in the air to be winnowed by the wind. On a breezeless day, he would puff with his breath at a shovelful of sand until he was blue in the face. But desert placers were hard to find and seldom paid well. Men who sought them were generally loners or veterans too old to compete in the hectic rush to a rich new strike.

In the better-watered gold country, the lucky prospectors who got onto the rich claims first would establish themselves in rough comfort, replacing their tents and lean-tos with log cabins, and doing very well as long as the gold lasted. The contemporary historian Hubert H. Bancroft, describing the productive placers on Idaho's Salmon River in 1861, wrote: "It was no uncommon thing to see on entering a miner's cabin a gold-washing pan measuring eight quarts full to the brim, or half filled, with gold dust washed out in one or two weeks." Since a full eight-quart pan held more than $2,500 worth of gold, the successful prospectors could well afford to indulge their taste for imported canned or bottled delicacies, which served a second purpose into the bargain. Said Bancroft, "All manner of vessels such as oyster cans or pickle bottles were in demand in which to store the precious dust."

As soon as a new-found placer had proved its worth, prospectors began looking for even greater wealth: the mother lode from which the river gold came. As a rule, there was a lode somewhere in the vicinity, though not always. For example only placer gold, no vein, was discovered in the Klondike. In any case, the search for a lode brought prospectors upstream from the placer, often for many miles. With patience and luck they would come upon what they called a blowup—the protruding end of an underground vein of cloudy, rust-stained quartz. The lucky finder might break off a piece of the rock and clap it to his tongue to remove dust and highlight the surface. If telltale yellow specks appeared, any normal prospector would jump for joy, or at least give a triumphant shout. If the specks did not appear, the prospector would pan some crumbled quartz for gold, or crush a chunk of rock to get a pannable sample. That was how miner John Gregory discovered that he had struck it rich in May 1859.

Gregory, a red-bearded mule skinner from Georgia, happened to be prospecting in the Colorado Rockies at the same time that George Jackson began working his placer. In fact, Gregory found his outcropping of quartz on the north fork of Clear Creek, just a few miles from Jackson across a long high ridge, and it was a much richer strike than Jackson's. Using pick and shovel, Gregory and his several partners hacked away at the weathered quartz, which miners called blossom rock or picture rock. A few feet down, Gregory's pick bit into softer material. He dug out some decomposed rock and dirt and raced with it to the creek. Then, with trembling hands, he panned the material in the icy water. When all the dross had been washed away, the pan glittered with four dollars worth of gold dust. Soon a couple of partners arrived in Denver with glittering samples, and the site of their strike was promptly dubbed Gregory Gulch.

By dint of brutal labor, Gregory and his friends pulverized tons of quartz and panned uncounted thousands of dollars worth of gold from their lode. Even so, they lacked the big capital needed to develop and exploit their mine, and no circumspect banker would trust them—unknown drifters—with large enough loans to finance the drilling of deep tunnels, the buying of machinery, the hiring of miners and the building of an ore-crushing stamping mill. Thus Gregory's group—and others like it—faced a choice between grinding labor and selling out their interests for a quick lump-sum profit. Most miners chose to take the best price they could get and to move on to the pleasanter work of looking for another strike. Gregory was paid $22,000 for his share. George Jackson, who made a second strike years later in Ouray, Colorado, sold that mine for $40,000. Little enough in both cases, but far more than many another prospector ever got for all his struggles.

Sometimes, the clue that led to new lodes was not a placer deposit but a piece of rock called float—a chunk broken off of the lode itself. Rainstorms, earthquakes and landslides would nudge float downslope, and the prospector who found a piece had a difficult task tracing his find back to its source. Lucky prospectors—and they were few—would find the path upslope to the lode marked with other bits of float. More commonly, the prospector would find no more float; instead, he would work his way uphill, panning samples of dirt along the way, and continue his course as long as the merest trace of color appeared in his pan. When the yel-

As a buddy looks on from a bunk, two miners enjoy a game of cards in their cabin in Colorado's Cripple Creek mining district in the 1890s. Early crude structures were replaced by such sawn-lumber buildings when an ore body proved rich and lasting and the region around it sprouted amenities.

low specks stopped showing, he would branch off laterally, panning more dirt and digging experimentally for a buried vein. This methodical approach might pay off, but it took a great deal of time. One Bob Womack ran tests for 12 full years before he finally discovered a rich vein in Cripple Creek, Colorado.

Womack, a lackadaisical drifter with a weakness for booze, picked up his piece of float in 1878 while working on a ranch at Cripple Creek, so-called because stray cattle were sometimes lamed in the boulder-strewn stream. The dull gray rock was nine inches long, three inches wide and unusually light for its size. Womack sent his rock to a friend in Denver, who had it assayed for him. The float proved out at $200 a ton —rich enough to be very exciting. Whereupon Womack launched his fitful long-term search for the vein from which the float had come. From his home in a wilderness cabin, he shambled drunkenly across the rugged landscape, digging holes for miles around, earning himself the nickname Crazy Bob. It was not until 1890 that his search ended successfully in Poverty Gulch, not far from where he had found his float.

When Womack announced his strike, his neighbors were inclined to laugh. Cripple Creek had long since been dismissed as a barren backwater. Knowledgeable prospectors and even professional geologists had climbed to an altitude of nearly 10,000 feet to investigate the mountain valley, and had found scant evidence of mineral wealth. No one ever dreamed that the valley lay atop an ancient volcano whose throat and radiating vents were choked with high-grade ore and veins of nearly pure gold. The local people found it hard to believe that anyone—least of all Bob Womack—could strike it rich at Cripple.

In the next year, Womack made a serious effort to prove he was right. He started digging a shaft for his mine, which he named the El Paso, and he eagerly displayed ore samples that assayed at $250 a ton. But his reputation as Crazy Bob discouraged men with money from investing the capital he needed to develop his mine. Meanwhile other men—notably a prospector named Marion Lankford and a moonlighting carpenter named W. S. Stratton—made strikes of their own nearby, and shipped out some $200,000 worth of ore.

By 1893, the gold rush at Cripple Creek was on in earnest. Some 10,000 miners and prospectors swarmed into the district that year, and they kept on coming at the rate of 500 a month. In 1900, the town of Cripple Creek had a population of 25,000 and its mines yielded $18 million in gold. Eventually, Womack's mine earned three million dollars, but not for its luckless discoverer. On a drinking spree, Crazy Bob sold the El Paso for $300 and moved away. He died broke in Colorado Springs in 1909.

Although most prospectors were as eager as Bob Womack to tell the world of their lucky strikes, others took great pains to conceal their good fortune and keep the gold for themselves. While George Jackson waited in Denver for spring weather, he mentioned his Clear Creek strike to only one trusted friend, whose mouth —Jackson wrote—"is as tight as a Number 4 beaver trap." Yet where gold was concerned, secrecy never worked; one way or another, the news always leaked out. Jackson himself had to break his silence and recruit partners to share expenses and labor. Any prospector who asked confidently for a grubstake was likely to arouse curiosity, and so was any stranger in town who spent a lot of money on tools and provisions.

The mining frontiers teemed with opportunists who were constantly on the prowl for information that they could cash in on quickly. Professional assayers had a moral obligation not to discuss prospectors' samples, but a good piece of picture rock, shot through with threads of gold, would loosen any man's tongue. Merchants, who were paid in raw gold, knew that dust and nuggets were distinctive in color, yellower or redder depending on the amount of silver or copper prevalent in their area of origin. And when a stranger showed up with gold that differed in tint from the local metal, the merchant might give him a thoughtful glance and casually ask him where he hailed from. If the prospector gave an honest answer, he might return to his diggings to find the place crawling with uninvited newcomers.

In the early days of the period, prospectors thought only of gold. But it became apparent that certain places on the Western mountains were prodigiously rich in silver as well. Though refined silver was worth one tenth of an equal weight in gold, an exceptional vein of silver-bearing ore might fetch $7,000 a ton—considerably more than the average gold-bearing ore.

However, silver, which combined far more readily than gold with other minerals, was harder to recognize

A guide to the miner's gritty argot

Amalgamation: a process using mercury to collect fine particles of gold or silver from pulverized ore. Both precious metals dissolve in the silvery liquid, while rock does not; they can later be released by applying heat or pressure to the mercury.

Bonanza: the discovery of an exceptionally rich vein of gold or silver.

Borrasca: an unproductive mine or claim; the opposite of a bonanza.

Claim: a parcel of land in a gold field that a person was legally entitled to mine because he had staked it out and recorded his title. The dimensions varied according to local custom.

Claim jumping: stealing someone else's mining property — usually after it had been staked out but before it had been officially recorded.

Colors: the particles of gold gleaming amid the residue in a prospector's pan after washing.

Coyoting: a method used by miners to reach gold deposits resting on bedrock without excavating all of the overlying soil. After a vertical shaft was sunk, tunnels radiating like wheel spokes were dug along the bedrock.

Crevicing: removing gold from the cracks and crannies of rocks by prying it out with a knife.

Cross-cut: a mine tunnel going across an ore vein, used for ventilation and communication between work areas.

Drift: a mine tunnel following the direction, or "drift," of a vein; opposite of a cross-cut.

Gallows frame: the wooden or steel scaffold at the top of a mine shaft carrying the hoisting rope.

Gangue: worthless minerals mixed in with valuable ore.

Giant powder: a miner's expression for dynamite.

Grubstaking: supplying a prospector with food and gear in return for a share of his findings.

Gumbo: the bane of the miner's existence — sticky wet clay.

Hard rock: ore that could be removed only by blasting, as opposed to ore that could be worked with hand tools.

High grading: the theft of chunks of ore by miners, who usually took only the valuable high-grade pieces.

Horse: barren rock interrupting a vein of ore.

Lode: a clearly defined vein of rich ore. The principal vein in a region was called the "mother lode."

Muck: the debris left after blasting hard rock. The miner who shoveled this ore-bearing material into a car or chute was known as a mucker.

Placer: a deposit of sand, dirt or clay, often in an active or ancient stream bed, containing fine particles of gold or silver, which could be mined by washing. The word is the Spanish for submarine plain, and rhymes with "passer."

Pyrite: fool's gold; a mineral composed of silicon and oxygen that is often mistaken for real gold.

Quartz: a crystalline mineral, often transparent, in which gold and silver veins were most commonly found.

Salting: planting rich ore samples in an unprofitable mine to attract unwary buyers.

Shaft: a vertical or inclined excavation; usually a mine's main entrance and hoistway leading to the tunnels where the ore was dug.

Sluice: a wooden trough for washing placer gold. As soil was shoveled into a steady stream of water, gold and other heavy particles sank to the bottom where they were caught by cleats, known as riffles. Some small, portable sluices, or rockers, could be rocked back and forth like a cradle to hasten the washing of gold.

Sourdough: an experienced prospector; traditionally one who had the foresight to save a wad of fermenting dough to leaven the following day's bread.

Stamp mill: a device that was powered by steam or water in which ores were pounded to a fine powder by heavy iron stamps, rising and falling like pile drivers.

Toplander: an aboveground worker at a mine.

Turned house: a mine tunnel that took a sudden change in direction.

Widow-maker: a compressed-air drill, used to bore holes for dynamite in hard rock. Prolonged inhalation of the fine dust created by early models of this drill subjected miners to a deadly lung disease called silicosis.

Winze: a passageway usually connecting two tunnels at different levels.

Freighters pull up at Swift's Station on the Kings Canyon
Toll Road between Lake Tahoe and Carson City, Nevada,
in the mid-1860s. This substantial caravansary near the
crest of the Sierra Nevada was a welcome stop for team-
sters who were hauling Tahoe pine to the Comstock Lode.

because it took on their various colorations. In the Comstock Lode, which yielded $105 million in silver in two decades, the white metal was combined with gold, and was dark blue in color. In parts of Colorado, ore containing silver and lead was as black as tar. Elsewhere, silver ore was yellow, white, pale green, dark red or brown.

To make silver hunting even trickier, there was no simple test for silver that a prospector could conveniently make under field conditions. The only sure way to identify silver ore and determine its richness was to test samples with nitric acid and hydrochloric acid, both of them dangerous to carry over rough terrain. As a result, prospectors were never certain that they had found enough silver to mine commercially until their ore samples had been assayed professionally. Moreover, silver seldom occurred in easily mined placers; big capital was required to dig out and refine the underground ore.

In spite of such difficulties, many prospectors made fabulous silver strikes. A classic example was Noah Kellogg, who stumbled upon veins of silver-bearing ore in the Coeur d'Alene district of northern Idaho. Kellogg was one of many prospectors who attributed their luck to some animal—burro or horse, badger or crow. The animal hero of Kellogg's find was a jackass.

The story was recorded by a prospector named Jim Wardner in a memoir immodestly entitled *Jim Wardner of Wardner, Idaho, by Himself.* According to Wardner, Kellogg and a few companions were prospecting in the hills when Kellogg's jackass wandered off and got lost, as it frequently did. "Next morning," Kellogg told Wardner, "we started out to find him. His tracks were plain, and now and then we found great wads of his hair where he had climbed over the down timber and scraped his sides against the logs. How under the heavens the little devil managed to get through that place I can't tell; but after we got into the canyon proper his trail was easy. Looking across the creek we saw the Jack standing upon the side of the hill, and apparently gazing intently across the canyon at some object which attracted his attention.

"We went up the slope after him, expecting that, as usual, he would give us a hard chase; but he never moved as we approached. His ears were set forward, his eyes were fixed upon some object, and he seemed

The bank where dust turned into $10 "mint drops"

"The little engine that drives the machinery was fired up, belts adjusted, and between three and four o'clock 'mint drops,' of the value of $10 each, began dropping into a tin pail with the most musical chink." Lest anyone think $10 a bit steep for a piece of candy, the *Rocky Mountain News* of July 25, 1860, hastened to explain that Denver's newest business was not a confectionery but a private mint, turning the gold dust of the region's mining camps into coins proudly stamped "Pikes Peak Gold." Its proprietors were Clark, Gruber & Co., who became the town's most respected bankers.

In the wake of the great 1859 gold rush, miners and merchants alike had discovered that "dust" had certain drawbacks as a medium of exchange. Most miners carried a pocket scale to measure out the price of everything from a drink to a pickaxe. But even so, disputes easily arose, with each party to a transaction suspecting the other — frequently with good reason — of cheating on quality or weight.

The problem eased a little with the arrival of the first bankers, who bought gold dust in exchange for hard currency, then shipped the dust back East for minting. But this was expensive and time-consuming, as Clark, Gruber & Co. had learned. Shippers levied a 5 per cent express charge each way, plus another 5 per cent for insurance coming and going. Moreover, the gold had to travel by stagecoach, and it might be in transit for as long as three months.

It occurred to E. E. Gruber that his bank should be coining its own

Denver's do-it-yourself mint occupied the Clark, Gruber bank's basement.

money. Was that legal? Austin M. Clark, an attorney and one of two brothers who were Gruber's partners, could find no law that forbade private minting, provided the gold was full-weight. Accordingly, Milton Clark set off for Philadelphia to purchase coining equipment, which reached Denver by ox-team four months later.

Their first coins, minted from gold dust that contained a natural silver alloy, were worth more than the government's. When this mintage proved too soft, the firm added alloy to the dust and still turned out a product purer than federal money.

After Colorado's formal recognition as a United States Territory in 1861, officials grew nervous about the propriety of a mint in their midst. Eventually, in 1863, Clark, Gruber & Co.'s property was purchased by the federal government. Although it had operated less than three years, the Denver mint had coined a total of $594,305 and had established an enviable reputation for probity and fair dealing. Its money had circulated everywhere in the world, turning up in Europe and even Australia. In a valedictory tribute, a Denver weekly paper avowed that "Messrs. Clark, Gruber & Co. have done more than any others to prove to the world the richness of our mining districts and the excellence of the gold produced."

wholly absorbed. Reaching his side, we were astounded to find the jackass standing on a great outcropping of mineralized vein-matter and looking in apparent amazement at the marvelous ore chute across the canyon, which then, as you now see it, was reflecting the sun's rays like a mirror. We lost no time in making our locations, and where the Jack stood we called the Bunker Hill, and the big chute we named the Sullivan (after a companion)."

The ore was galena, a mixture of lead and silver, and the Bunker Hill and Sullivan mines turned out to be bonanzas. Within a month of Kellogg's discovery, a bustling town, later named Wardner, sprang into being near the site; its mines were to yield $300 million in the next 60 years. The tale of Kellogg's jackass enjoyed a long life, despite the efforts of sober-minded men to put an end to it. Thomas A. Rickard, a brilliant mining engineer of the period, pointed out what plumbers and housewives knew full well: that silver and lead soon lose their brightness and turn drab gray when exposed to the air. Wrote Rickard, "The talk of a glittering mass of silvery ore sticking out of the mountain-side so brilliantly as to mesmerize the ass, and others not any wiser, is pure moonshine."

There was a sad epitaph to the story. Some months after the great find, Kellogg wearied of searching for his perpetually wandering animal, tied some dynamite to its back and drove it off down a mountainside where not even a goat could keep its footing. After a while, a distant explosion told him he would never again have to chase that ornery jackass.

Whether gold or silver triggered a rush, the camps that mushroomed around the strikes all tended to follow the same general pattern. Most miners were willing enough to respect the rights and boundaries of men who made original claims. But as congestion mounted and claims were bought and sold, communities found it essential to set up some kind of ad hoc government to police the action until officials from the national or territorial government arrived. The results were thousands of fascinating experiments in grass-roots democracy.

Someone would decide that the time had come for a general meeting. As soon as the call went out, all the miners in a camp dropped their pans and pickaxes and assembled in a central location. They elected a pre-

siding officer, a recorder and perhaps a town marshal. They determined the logical boundaries of the gold field and pronounced the area a duly established mining district. Then, borrowing freely from the original customs of the forty-niners, they agreed on rules concerning claims. These rules were earnestly calculated to give every miner a fair chance.

With the exception of the discoverer of a new gold field a miner could stake only one claim in the district. The discoverer got two. In very rich ground, the basic claim size might be as small as 10 feet square; while in leaner areas, it might be as large as 100 feet. On narrow creeks, the placer claim might extend from bank to bank; on wider watercourses, it might go only to the thread or center of the channel, with room for access and work on one bank. Every claim had to be boldly marked, and the date, location and claimant's name had to be filed with the district recorder, who was often a saloonkeeper or merchant. Clarity, not elegance, was all that was required in marking the claim stakes. For example, one early claim notice warned all comers, "CLAME NOTISE—Jim Brown of Missoury takes this ground; jumpers will be shot."

Having staked a claim, a man was obliged to work on it—typically, at least one day in three—to retain title. If he did not show up for 10 days or two weeks, the claim could be taken by someone else. No man could dump waste earth or rock onto his neighbor's claim or interfere with the general water supply. A placer miner had free use of the river as it passed through his claim, but—in one rule that was often flouted —could not divert or impound the flow at the expense of his downstream colleagues. Naturally, claims could be bought and sold at will, but to prevent fraud or coercion, each transaction had to be witnessed by at least two and sometimes five disinterested men.

Claim jumping, theft and murder were the only crimes that seriously disturbed the miners. In all such cases, a meeting was speedily convened to deal with the suspect. Judge, jury, prosecutor and defender were elected (everyone in camp, including youngsters of only 15 or 16, could vote), and the trial went on at once. If the suspect was found guilty he received some sentence that could be executed without delay; summary punishments were the rule because new mining camps had few jails—no one wanted to build, pay for or guard

A wagon train bumps up boulder-strewn Bridge Street of year-old Helena, Montana, in 1865; another hazard in such streets came from prospect holes left unfilled. Helena was originally known as Last Chance Gulch, because prospectors hit pay dirt just when they were about to quit.

them. Mutilation—the cropping or removal of an ear or two—was sometimes recommended. Flogging was common, as was banishment; men were obliged to leave the mining district immediately and to remain out of it on pain of violence or death. Hanging was the routine sentence not only for murder but for theft of a particularly obnoxious kind, such as the stealing of scarce food in the Klondike.

The decisions of miners' courts, made on a moment's notice by wilderness amateurs, sometimes miscarried tragically. One such case, in a Colorado gold camp, involved a misanthropic miner who regularly got liquored up and took pot shots with his revolver at every passerby. When the miners' court convened to punish the drunken marksman, the consensus was that only banishment would prevent him from committing murder sooner or later. But then one miner suggested an insane alternative that was quickly adopted and carried out: a "preventive" hanging.

For the most part, however, common sense saved the miners from horrendous mistakes, and their decisions—especially in regard to claims—were generally approved in later years by proper judges and official courts. Indeed many judges expressed wistful admiration for the speed and directness of the old verdicts. James Wickersham, who was assigned to the territory of Alaska as a federal circuit judge in 1900, wrote with relish of the case of a dance-hall fiddler who had seduced a girl in a backwoods camp, left her pregnant and refused to marry her. A miners' meeting quickly resolved "that the defendant pay the plaintiff's hospital bill, $500, and pay the plaintiff $500, and marry her as he promised to do, and that he have until 5 o'clock this afternoon to obey this order; and resolved further, that this meeting do now adjourn until 5 o'clock."

The wretched fiddler conjured up lurid pictures of what the miners might decide to do at 5 o'clock, so he married the girl forthwith. "It would have taken my court two years," wrote Judge Wickersham, "with many pleadings, hearings and arguments, to give judgment, which in all probability would have been reversed on some technicality."

While the miners were hammering out their makeshift laws, they were joined and almost overwhelmed by an avalanche of other citizens who were not so much interested in getting gold out of the ground as in getting

it out of the miners themselves. Among the first nonmining arrivals were the wagon freighters, who opened supply lines and helped the camps expand. With them came merchants and saloonkeepers who went into business at once, pitching their tents wherever convenient and using rough planks laid between barrels as counters. Then came the professional gamblers, also using planks for their places of business.

One gambler, a man about 60 years old, was especially noteworthy because almost every denizen of the early mining camps was conspicuously young; one woman said she spent several years in Colorado before she saw a man with gray hair. This particular gambler had a full head of white hair, and he showed up early in the rough-and-tumble life of Deadwood, Dakota Territory, chanting a siren song that made his age profitable: "Come on up, boys, and put your money down—everybody beats the old man—the girls beat the old man—the boys beat the old man—forty years a gambler—the old fool—everybody beats the old man—put your money down, boys, and beat the old man."

Soon after the gamblers arrived, madams and their whores would roll into town in stagecoaches or even private carriages, amid raucous shouts of delight from the miners. Then the real-estate speculators would turn up, sometimes with a surveyor, and invariably with a carpetbag full of plans for laying out the perfect city and selling lots at lofty prices. And always gold, or rumors of gold, attracted professional criminals. Few were daring crooks of the sort that was glamorized in the Eastern press. Most were small-time swindlers or sneak thieves and thugs—men whose notion of a good day's work was to slug a miner in a dark alley and make off with his poke.

For the first several months of a gold camp's existence, it was likely to resemble Custer City, Dakota, as that wide spot in the road was described by a traveler in 1876: "There were but few houses completed, but many under construction. The people were camped all around in wagons, tents and temporary brush houses. The principal business houses were saloons, gambling houses and dance halls, two or three so-called stores with very small stocks of general merchandise and little provisions." If the diggings continued to produce gold or silver, the camps would grow fantastically in size and even change in character. In 1879, when lusty Leadville was two years old, it was a boisterous mecca for 20,000 miners, replete with 82 saloons and 35 houses of prostitution; yet already the first signs of stability had appeared there—a public school and seven churches.

However, most mining camps did not survive long enough to become solid communities. All too soon, the precious metal that gave them life began to dwindle, and with it the towns' population. Placer gold was reduced to a point where prospectors, barely able to eke out a living, moved on to a newer, more promising strike. Veins of gold or silver slowly petered out or suddenly disappeared, broken or buried by some colossal earth convulsion ages before. Many a mining town, left with no source of income, became a ghost town.

Yet the persistence of mining in the West was nothing short of phenomenal. Some mining districts enjoyed as many lives as a tomcat, with new strikes being made as old ones played out. In several areas, silver was found after the gold was depleted; in turn persistent miners left their waning veins of silver to find a galaxy of industrial minerals.

The Leadville district of Colorado, located 10,000 feet high in a valley of the Rockies, was a perfect ex-

Amid empty cans, Colorado prospectors break for dinner in the 1880s. Beans, flapjacks and sourdough bread were the miners' staples.

ample of this evolution in Western mining. In 1860, a prospector named Abe Lee struck gold there in a mountain stream flowing from the Mosquito range. The stream gorge, named California Gulch, instantly became the site of a rip-roaring gold rush, with crude dwellings strung out along the narrow gulch for a mile. But two years later, the gold ran out and miners abandoned their long skinny town in droves. For all practical purposes, the area died as a mining district—its first death. It stayed dead for 13 years.

The district's second life began in 1875 when Will Stevens, a prospector with a lively imagination, arrived in California Gulch to give the old depleted diggings a new try. Stevens was curious about the black sand and rock that underlay the area and finally had a sample assayed. It proved to be a lode of carbonate of lead containing two and a half pounds of silver to the ton. Naturally Stevens and his friends bought up all the idle claims they could.

When news leaked out in 1877, an instant boom-town mushroomed at the foot of the gulch, and miners named the place Leadville in honor of the silver-bearing carbonate of lead. The area proved so rich that one mine produced silver ore worth $118,500 in a 24-hour period. By 1880, $12 million in silver had been shipped out of the district.

Soon thereafter, Leadville's silver waned, and the town seemed headed for oblivion again. But Leadville got a new lease on life in the 1880s when miners struck new deposits of mineral wealth. The minerals were not nearly so noble as gold and silver, but they were increasingly in demand for new manufactured products: copper and zinc. Leadville prospered.

But gold was the glamor metal of the pioneer epoch, the main chance and chief attraction for every man who ventured forth in hopes of striking it rich. Fittingly enough, the most sensational find of the period—and perhaps of all time—involved gold and was made in 1914 in Colorado, where the epoch began a half century earlier. It was not a placer; by then most of the loose gold and nuggets that could be mined economically had been scraped from the stream beds of the West. By then, too, the Colorado miners were digging far down into hard rock, following veins that ran hundreds and even thousands of feet below the surface. The climactic find was made in one such mine, known

as the Cresson and located in Cripple Creek, not far from the place where booze-loving Bob Womack had made his strike in 1890. The treasure was unbelievable, and the few men who saw it first hardly knew what to do about it.

Dick Roelofs, a mining engineer whom the Cresson's absentee owners had left in charge of operations, had been working with a mining crew in a tunnel 1,200 feet underground. They were cutting their way along a good lode when they suddenly holed through hard rock into a cavity, technically known as a geode but usually called a vug in miners' lingo. This was the biggest, richest vug Roelofs had ever heard of, and the discovery threw him into a tizzy. After glancing through the aperture with his miner's lamp, he sent his men away to fetch an ironworker, who installed a vault door barring the way into the vug. Roelofs posted armed guards at the door.

To keep everything aboveboard, Roelofs then rounded up two witnesses—substantial men who knew mining and whose word was certain to be trusted by the owners. The trio descended the Cresson's shaft. Roelofs led the way to the steel door and opened it.

The men stepped into a wonderland of solid gold. The vug was 20 feet long, 15 feet wide and 40 feet high. Its walls blazed with millions of gold crystals and 24-carat flakes as big as a thumbnail. Pure gold boulders littered the floor amid piles of white quartzlike spun glass. It was dreamlike wealth and beauty such as this that drove miners to delve into the earth.

One of Roelofs' witnesses hazarded a hoarse guess: the wall surfaces alone might be worth $100,000. He had underestimated by more than 300 per cent. After Roelofs' most trusted miners had scraped the walls and filled 1,400 sacks with crystals and flakes, the lot sold for $378,000. Lower grade ore—1,000 sacks of it—brought $90,000 more. Then the crew mined the outer layers of rock—ore worth more than $700,000. The vug was stripped in four weeks, and it enriched the mineowners by the sum of $1,200,000.

Dick Roelofs, too, profited no little from the Cresson vug. He soon became known from coast to coast as "the miracle miner," and was made one of the major stockholders in the Cresson mine. Fame and wealth permitted Roelofs to put first things first: he removed East to New York to enjoy women and wine.

The gold rush that overran the Sioux's sacred domain

As early as 1833, scattered parties of white men who had heard Indian legends about gold in the Black Hills of Dakota ventured into that granitic region to reap the riches. Their troubled passage was marked by the rusted remnants of shovels, skulls with silver spectacle frames — and such testimony as an incised slab of stone bearing the words, "Got all the gold we could carry. Nothing to eat and Indians hunting me."

Then in 1868, the Black Hills were closed to all white men, ceded by treaty to the Sioux as their perpetual sacred wilderness. However, in 1874, the Army sent Lieutenant Colonel George Custer to find a site for a fort in the hills. Custer's entourage included two experienced prospectors, and when they found what the headline-loving colonel tantalizingly described as "gold in paying quantities," a ragtag invasion

of fortune seekers began. Hastily, the government tried to bargain the hills back from the Sioux. The Indians instead went on a rampage.

In the end, the Indians were forced to yield, and the prospectors overran the hills: by 1877, there were some 7,000 scrabbling in the creeks and gulches and — according to one account — gold worth thousands of dollars was stored in pans "like so much old iron."

45

U.S. troops pass a stockade built by prospectors who invaded the Black Hills after Custer confirmed the presence of gold. The Army evicted such trespassers until negotiations for purchase of the hills broke down; thereafter, the government ceased trying to maintain Indians' treaty rights.

47

Teton Sioux make camp near the 6,000-square-mile Black Hills region, believed to be an abode of the spirits. After refusing a six-million-dollar offer by the government, they had to surrender the hills for nothing in the fall of 1876 in order to get their rightful allotment of winter provisions.

49

Rude cabins, tents and hybrid shelters distinguish Gayville as a newborn Black Hills community in early 1876. Founded by miner Alfred Gay in gold-rich Deadwood Gulch, the town quickly expanded to merge with nearby settlements; within a year, their combined population was 3,000.

ROATH
Herrmann & Treber.
SALE LIQUOR DEALERS
St. Louis Lager Beer
DENTIST.
BANK.
STEBBINS, POST & CO.
JOB PRINTING

Deadwood, the capital of Black Hills gold, teems with traffic in what a resident called "mud of adhesive properties rare, its depth unfathomable." Here the banks handled $100,000 a day, a bunk cost $1 a night and "Coal Oil Johnny" monopolized kerosene sales at $3.75 a gallon.

2 | A lode to outshine King Solomon's mines

Of all the names that have been attached to metal-bearing earth, easily the most stimulating to the imagination of men in the latter half of the 19th Century was the "Comstock Lode." Not even the Biblical King Solomon's mines could conjure up such visions of wealth incarnate. And it was all true: one deposit on the eastern slope of the Sierra Nevada, named after one of its first claimants (*below*), embraced upward of a third of a billion dollars worth of precious metal, most of it silver.

To place a value on the barren mountainside that held the precious metals, speculators priced Comstock mines by the foot. The lode's fabulous Ophir, for example, was once quoted at $4,000 per foot along a claim nearly a quarter of a mile in length. And for years, the values seemed to go nowhere but up. "Feet that went begging yesterday were worth a brick house apiece today," wrote Mark Twain in 1871. Twain had been working as a reporter on the Virginia City *Territorial Enterprise*, the area's leading newspaper, and he could scarcely believe his surroundings: "Think of a city," he said, "with not one solitary poor man in it."

Henry Comstock, for whom the lode was named, sits at far left in this painting of the discoverers staking the first claim in June 1859.

55

Millions in silver from bothersome blue muck

"Precious metal" meant only one thing in the decade following the discovery at Sutter's mill. It meant gold. Very few men thought of silver, and when silver was at last found, prospectors did not recognize it and shoveled it aside. The discoverers of the $400 million Comstock Lode, probably the greatest single mineral strike in history, were not only *not* looking for silver but regarded its ore as yet another obstacle provided by fate to prevent poor lads from making an honest living.

The discovery was a long time coming, considering that so many people had passed so close to the Comstock. Thousands of forty-niners and other emigrants traveled to California along a trail that led west from Salt Lake City across alkaline deserts to the eastern slope of the Sierra Nevada. They paused for a few days in the Carson River valley below Lake Tahoe to refresh themselves before starting the tedious climb over the mountains to the promised land. While recovering their strength, some of the emigrants took to prospecting on the side of Mount Davidson, which loomed over the valley. As early as 1850 they found specks of gold on the banks of a creek that flowed down the mountain into the Carson. The quantity of metal was small and the creek bed lay in the bottom of a shallow gulch on Davidson's southern slope. Nonetheless, before pushing on to what they were sure would be better diggings in California, the emigrants gave the gulch the high-flown name of Gold Canyon.

On Davidson's northern flank there was a similar gulch called Six-Mile Canyon. Small showings of gold had been found there as well. The two gulches approached each other as they extended up the mountain, and at their terminals they were about a mile apart. Along a line between these terminals, lying roughly north-south at an altitude of 6,400 feet, was the Comstock Lode. No place on earth was richer in silver, and there were vast amounts of gold as well. The lode was traceable on the surface by scattered quartz outcroppings, but they showed only faint hints of the wealth underground. The fabulous ore was hidden, although in places it was so shallowly buried that a man could almost scuff it up with his boots. A willing worker with a pick could — and would — make himself a fortune.

But in the early 1850s, Gold Canyon did not attract energetic, ambitious miners. Those who scratched out a living there were easygoing — content with slim pickings. With modest effort they earned perhaps three to four dollars a day, enough to pay for food, supplies and a good drunk on Saturday night, but little more. By the mid-1850s about 100 of them had small placer claims along the creek, which they worked with pans, rockers and gold-sifting troughs called long toms. In the winter most of them lived in shelters made of rocks plastered with mud, and in the summer in brush huts inferior even to the dwellings of the Washoe Indians for whom the region was named.

Among the residents of the Mount Davidson gulches in 1859 were some picturesque citizens. One was a Canadian-born blowhard named Henry T. P. Comstock, nicknamed Old Pancake because, it was said, he was too lazy to bake bread and consumed his flour in the form of flapjacks. Another was James Fennimore, who seems to have been involved in a murder in California, hastily departed for Washoe, and changed his last name to Finney. He was commonly called Old Virginny after his natal state. There were also two Irish immigrants named Patrick McLaughlin and Peter

---

Just beyond the protection of timbers, a Comstock miner feels his way into the portal of a new shaft. Cautious entry was advisable, for the Nevada lode's ore and surrounding clays were terribly unstable.

O'Riley, not long off the boat, gullible, and easy marks.

By the early spring of 1859 Old Virginny Finney had worked his way up to the head of Gold Canyon and found some promising ground on top of a knoll. Others, including Comstock, joined him, staked claims beside his, and named the knoll Gold Hill. All of the claims—50 by 400 feet per man, following local rules—were for placer mining. None of the men suspected that they had struck a lode—the southern end of the Comstock, of which the knoll was an outcropping. When Old Virginny and Old Pancake scratched the surface they began to take out about five dollars per man per day, and when they dug down a foot or two they got $20 apiece. That was very good money—enough to warrant building some log houses and the foundation of a tiny town, also called Gold Hill.

Patrick McLaughlin and Peter O'Riley soon came to Gold Hill but, finding that all the good ground had been staked, hiked over to the head of Six-Mile Can-

yon and began to dig there, washing their dirt in a rill that ran down the gulch. The trickle was inadequate, so they decided to dig a small reservoir. At a depth of about four feet they struck a layer of blue-black material, quite different from the yellowish sand and clay in which they had been working. When they washed a few pecks of the stuff they found the bottom of their rocker covered with pale metal dust. It was gold, they were sure, but something was wrong with it. It was too light in color, apparently because of adulteration with some base metal whose nature they could not guess. Still, it was indisputably gold and there seemed to be a great quantity of it. Unknowingly, they had hit the Comstock near its northern end.

O'Riley and McLaughlin rejoiced, but not for long. Out of nowhere appeared Henry T. P. Comstock, who examined the pale dust and the blue-black sand and raised his bullying voice. The land was his, cried Old Pancake. He was also part owner of the spring

The hoisting works of the Yellow Jacket, Imperial, Central and Empire mines vie for room in a particularly rich section of the Comstock.

from which flowed the trickle of water, having earlier staked out 160 acres around it for a ranch. These were whopping, indecent lies, but Old Pancake would not permit the Irishmen to continue digging unless they took him and a friend of his, one Emanuel Penrod, into partnership. The Irishmen meekly agreed.

For the next fortnight, Comstock, Penrod, O'Riley and McLaughlin dug into the bluish sand and washed it to extract the pale metal with which it was flecked. Because it was contaminated (with silver, though no one as yet realized this), the gold fetched only eight or nine dollars an ounce when they sold it. That was just half the price commanded by the pure California product. Still, the men were making hundreds of dollars a day. The sole drawback in the operation was that the heavy blue sand clogged the rockers and made their arms ache from scooping it up and throwing it away.

None of the miners was curious enough about the sand to try to find out what it was. However, there were men of more inquiring mind in the area. After a few weeks a settler on the Truckee River gathered a sack of the "blue stuff" and carried it over the Sierra to an assayer in Grass Valley, California.

The assayer, a steady and responsible fellow, could not believe the results of his tests. So he repeated the tests with the utmost care—and got the same answer. The sample went $3,000 in silver and $876 in gold to the ton. In giving its dollar value per ton, he was merely following standard practice in the mining country. He had no way of knowing whether there might be 10 tons, a million tons or only a few ounces. All he was saying was that if a man happened to have a ton of it—in heavy Comstock ore, that would have been a cube less than two and a half feet on a side—it would be worth $3,876.

Now, however honest and professionally close-mouthed the assayer may have been, $3,876 was not a figure that could have been kept quiet for long. In-

But the Bullion stands alone—for good reason: though it was at the center of the lode and went down 2,725 feet, the mine was barren.

deed, before breakfast the next morning the foremost citizens of Grass Valley, with the local judge running first, were frantically racing over the mountains to Washoe, anxious to stake claims for themselves or, failing that, to buy up the claims of the ignorant miners before the miners heard about the $3,876. But when the Californians got to the diggings they found that the miners had already put out a forest of stakes.

Old Pancake, Old Virginny and the others may not have known that the blue stuff was silver, but they did know it contained a lot of gold. (The eventual output of all the mines on the Comstock was 55 per cent silver and 45 per cent gold.) Moreover, they now realized that they were not working placer claims but lode claims, which entitled them to pre-empt more land.

A placer claim was finite, 50 by 400 feet, but a lode claim was indefinite. On the Comstock, each man was allowed 300 feet along the north-south lode, plus all the ground east and west of his claim that the various offshoots of the lode might occupy. By old custom, an additional 300 feet along the lode was granted

jointly to its discoverers. All told, O'Riley, McLaughlin, Comstock and Penrod had 1,500 feet, which they called the Ophir mine after the source of King Solomon's gold mentioned in the Old Testament.

The men were not notable Bible readers. Ophir merely happened to be a favorite name for mines in the Old West. Old Virginny Finney was not a student of the Good Book, either, but he knew about baptism. One night while wandering drunk along the Ophir line, he dropped his whiskey bottle and broke it. Trying to make the best of the catastrophe, he is said to have bawled, "I christen this ground Virginia!" Such, at least, is the explanation offered to those who wonder why the queen metropolis of the Comstock in westernmost Nevada is called Virginia City.

Thus it happened that the front-running judge and the other smart Californians arriving on the scene found that the ground had been staked all the way from Old Virginny's claim at Gold Hill on the south to the Ophir on the north. The newcomers had no choice but to buy out the original claimants. As it developed, they sold out for what came to seem rather trifling sums. But there was another factor: the characteristic rootlessness of the prospector. It made him prefer a quick, certain return to one that looked risky or remote. The abilities of the prospector and the abilities of the developer were rarely found in one man; in the history of Western mining the persistent reality was that the prospector who found precious metal also lost it—while the developer moved in and kept it.

As the smart operators arrived, Henry T. P. Comstock briefly retained his bully-gotten share of the Ophir, a mine that ultimately yielded more than $11 million. For weeks he wandered along the lode, boasting so loudly that his name became attached to the Comstock as discoverer and sole proprietor. He then sold his share to James Walsh, the speedy judge from Grass Valley, for $11,000. Comstock later invested his $11,000 in a supply store and lost it. He continued to believe, however, that he owned the lode and the town that had grown up around it, and with the mad grandiloquence of a King Lear announced that he was willing to allow his "tenants" to live in Virginia City rent free, "for the winters are cold, and the people poor, and their need is greater than mine." Eleven years after the discovery of the lode, the demented, drifting Comstock shot himself and was buried without a headstone in Bozeman, Montana.

Comstock's friend Penrod sold his interest in the Ophir for $5,500 and got $3,000 for another claim he held on the lode. He then faded into total obscurity.

Old Virginny Finney disposed of his holdings for an uncertain sum, although it must surely have been more than the "old horse, a pair of blankets and a bottle of whisky" that tradition assigns him. Whatever the case, he remained drunk for a year or two, fell off a horse and died of a fractured skull. His friends gave him a fitting funeral and passed a resolution proclaiming him "a generous, charitable and honest man to whom more than any other they are indebted for the discovery of the mineral wealth of this Territory."

Of the Irishmen, Patrick McLaughlin sold his share in the Ophir for $3,500. He became a cook for other miners and died a pauper. Peter O'Riley was the most tenacious of all. He held out and eventually got about $45,000, but he did not live to enjoy it. He soon began to hear the voices of spirits, telling him to dig in a mountain nearby. One observer reported, "As he wielded pick and sledge, their voices came to him out of the darkness which walled in the light of his solitary candle, cheering him on. He was sent to a private hospital for the insane and in a year or two died there, the spirits to the last lingering about him."

The newcomers to the Comstock heard no spiritual voices. They were tough, single-minded men who had but one idea: get rich. They were going to do it, moreover, in a Godforsaken place more than a mile up the side of a mountain so barren that all it could support was sagebrush and stunted pines. To the north, east and south, the mountain was surrounded by deserts; to the west, there was the barrier of the snowy Sierra. There was little water. There were no roads. Every can of beans and jug of whiskey, every plank and tent, would have to be dragged to the Comstock over trails that snaked through canyons, over high passes or wound along precipices. The effort would require as much energy, faith, skill and courage as Americans had ever mustered for a single industrial project.

Among the first of the aggressive developers to reach the Comstock was George Hearst, whose son was to become a famous publisher. The elder Hearst, son of a Missouri farmer, had been in California since 1850,

After making his first fortune as part owner of the Ophir, George Hearst progressed through a spectacular career as a mining promoter, publisher, philanthropist and then a United States Senator from California.

working unsuccessfully as a placer miner and then as a storekeeper. He returned to his first love, mining, after locating two promising quartz claims. But when he heard the news from Washoe in 1859, he dropped his pick and hurried across the Sierra. Hearst quickly made a deal with Patrick McLaughlin to acquire his one-sixth interest in the Ophir mine for $3,000, dashed back to California to raise the money, then returned with friends to the Comstock and started digging.

With great labor, working against the approach of winter, Hearst and his companions managed in about two months to extract 38 tons of high-grade ore from the Ophir. They loaded it on mules and somehow got it over the mountain passes, then nearly blocked by snow, to a smelter in San Francisco.

The costs of transportation ($140 a ton) and smelting ($412 a ton) were enormous—but so was the yield of the ore, which came to $3,000 a ton and left Hearst and his friends a profit of more than $90,000. The gleaming white bars of silver bullion were paraded through the streets to a bank, where they were stacked in the window. San Franciscans, who had often been deceived by false or inflated stories of mineral strikes somewhere out yonder, were convinced that El Dorado itself lay just beyond the Sierra. Even though snow still lay in the passes many of them headed for the lode; ultimately, some 17,000 claims were filed, most of them worthless.

Hearst did not remain long on the Comstock. He sold his share of the Ophir at a large profit and went on to make a huge fortune at the Homestake gold mine in South Dakota and the Anaconda copper mine in Montana (page 114). But it was on the Comstock that he got his start.

The early excavations on the lode, notably at the Ophir and at Gold Hill where the ore bodies were close to the surface, were at first simply open pits. But soon the sides began to slump, and attempts were made to dig timbered shafts. Problems then arose because of the strange geological formation of the Comstock. It was not a single, clearly defined vein of ore, but a miner's nightmare. Ore was found in bodies scattered like raisins in a cake. Between and surrounding these bonanzas were walls of barren rock, belts of partly decomposed porphyry and sheets of clay. There were also underground reservoirs of hot water. The lode was unstable, likely to move or cave in when unsupported.

A more pleasing peculiarity of the Comstock was the astonishing size of the ore bodies. Elsewhere in the world, a vein of silver ore three feet wide was considered large. On the Comstock, there were deposits hundreds of feet wide, and much of the ore was so crumbly that a miner could hack it out with a pick.

The first Ophir shaft was sunk on an incline to follow the apparent tilt of the lode. As digging reached a depth of 180 feet, the ore body, which had been mea-

A MOUNTAIN CROSSHATCHED WITH CLAIMS

In this 1878 map of the lode, more than 50 major mining properties run in a generally north-south direction over a span of about eight miles. At top is a cross section of the Sutro Tunnel, intended to drain troublesome water from major ore bodies. In the plan view, the four-mile tunnel is represented by the heavy black line running up to the center of the lode. Virginia City stood on the mine sites concentrated at right center on the chart; Ophir, the original bonanza mine, was sunk in the very middle of the metropolis.

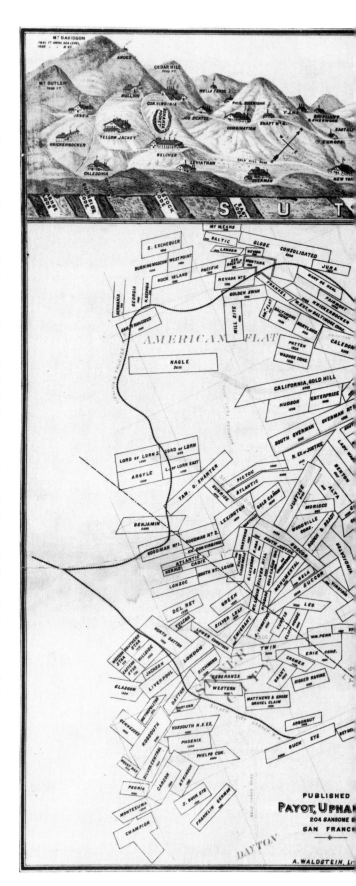

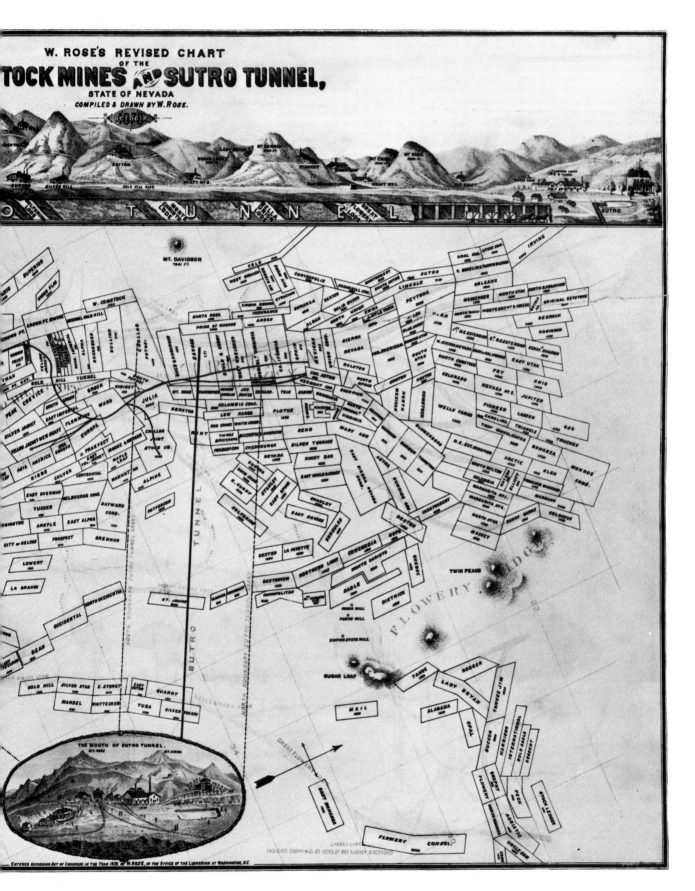

sured in inches near the surface, had broadened to 40 or 50 feet—and there was no safe way, then known, of working it. Although a fortune was in sight, miners would have been crushed to death if they dared touch it. The ordinary post-and-cap method of timbering—in which each set of timbers resembled a doorframe—was of no avail in an excavation 40 or more feet in height. Nor was it possible to use the so-called room-and-pillar system, in which columns of ore are left to hold up the ceiling. Comstock ore could scarcely hold up itself. What was needed was extraordinarily strong support—in three dimensions—capable of being extended in any direction as the ore was taken out.

To solve the problem the Comstockers called in a young consultant named Philipp Deidesheimer, a graduate of the Freiberg School of Mines in Germany, who had been practicing his profession in California for several years. At the time, Freiberg was the world's foremost institution of mining technology, a science that Americans were just beginning to grasp.

After several weeks of study, Deidesheimer developed a system of timbering in "square sets," at once so novel and so effective that it became famous overnight. Deidesheimer prescribed short, massive timbers, about 14 inches square and six feet long, mortised and tenoned at the ends so they could be assembled to form strong hollow cubes. Each cube could be interlocked with the next, as occasion demanded, in a manner somewhat reminiscent of a honeycomb. Square sets, devised in 1860, made possible the opening of the Comstock to great depths, and within a few years there were nearly a score of mines spread out along the lode and more than 100 in the general vicinity.

The ore bodies in the Comstock numbered about 30. They were all irregular in shape but, in arbitrarily round figures, they ranged in size from little pockets only 100 by 80 by 40 feet, upward to the aptly named Big Bonanza, which was at least 600 by 400 by 70. A half dozen of the ore bodies touched or were very close to the surface; the others were deeply buried. The method of exploring for these deep-lying bodies was the long-established one: vertical shafts, from which branched horizontal tunnels, or drifts.

The probing of the Comstock continued from 1860 to the late 1880s, but all of the major discoveries were made early—the Big Bonanza coming in 1873.

The mines yielded metal worth $1 million in 1860, $2.27 million in 1861, $6.24 million in 1862, $12.48 million in 1863, $15.79 million in 1864. There were periods of fall-off and resurgence, with a peak of $38.57 million in 1876, when the richest part of the Big Bonanza was hoisted to the surface.

Annual figures can be cited indefinitely, but the gold and silver that came out of the Comstock—even the grand total of nearly $400 million—at last loses its power to impress. A pile of bullion as big as a modern freight car, which is what it amounted to, is simply a very large pile and no great shakes in comparison with the mind of man. The real wonders of the Comstock were in the audacity and ingenuity it inspired. Deidesheimer's brilliantly conceived square sets were only the beginning. At once a new challenge appeared: where could one find an unlimited supply of timbers 14 inches square? In total the mines required about 80 million board feet of timber and lumber every year, as well as 250,000 cords of wood to fuel their numerous and very large steam engines.

The nearest source of wood was in the pine forests of the Sierra, and before long, the slopes were stripped of trees for a stretch of almost 100 miles. At first the wood was carried to the mines in huge wagons along a newly built road. But as the lumbermen were obliged to work farther and farther away, one of their number, J. W. Haines, dreamed up the V-flume, an invention that was soon to become standard equipment in Western logging. The V-flume (so called from its shape) floated timber and firewood for miles down the mountains to a point close to the mines.

The hauling in of machinery and other supplies, and the hauling out of ore, was more than mule- or horse-drawn wagons could economically accomplish. In 1869 a railroad was built from Virginia City to Carson City on the river, and later extended north to connect with the Central Pacific at Reno. The Virginia City-Carson City segment, 21 miles long, was blasted through solid rock for the greater part of its distance. It was all downhill (or uphill), and may well have been the crookedest railroad ever constructed. The 21 miles included 6,120 degrees of curvature, the equivalent of 17 complete circles, and a descent (or ascent) of 1,600 feet. There were innumerable deep cuts and trestles, and six tunnels with a total length of about half a mile. It took a master surveyor only 30 days to lay out the road, but it took nine months to build it with a crew of 1,200 men, most of them Chinese.

Most supplies hauled to Virginia City went into the mines, but there was a substantial tonnage of necessities for workers and their families. The populations of Gold Hill and Virginia City, which were in effect one town divided by a ridge, totaled about 25,000 at their peak in the 1870s. Most of the citizenry could afford to live reasonably well. The miners had formed a union as early as 1866; the next year they established an

eight-hour shift with a minimum wage of four dollars a day for underground work — a rate perhaps better than any in American industry at the time.

In response to the ample supply of spending money, the railroad fetched fancy furniture, carved woodwork and hewn stone for an opera house and for the six-story International Hotel, which was regarded by residents of Virginia City as the most luxurious establishment of its kind between Chicago and San Francisco.

Yet the life of the Comstock miners was anything but easy. They lived in unpainted shacks and boarding

A stage pulls into a tollhouse near the Comstock Lode in 1866. From Sacramento to Virginia City, a four-horse rig paid some $15 in tolls.

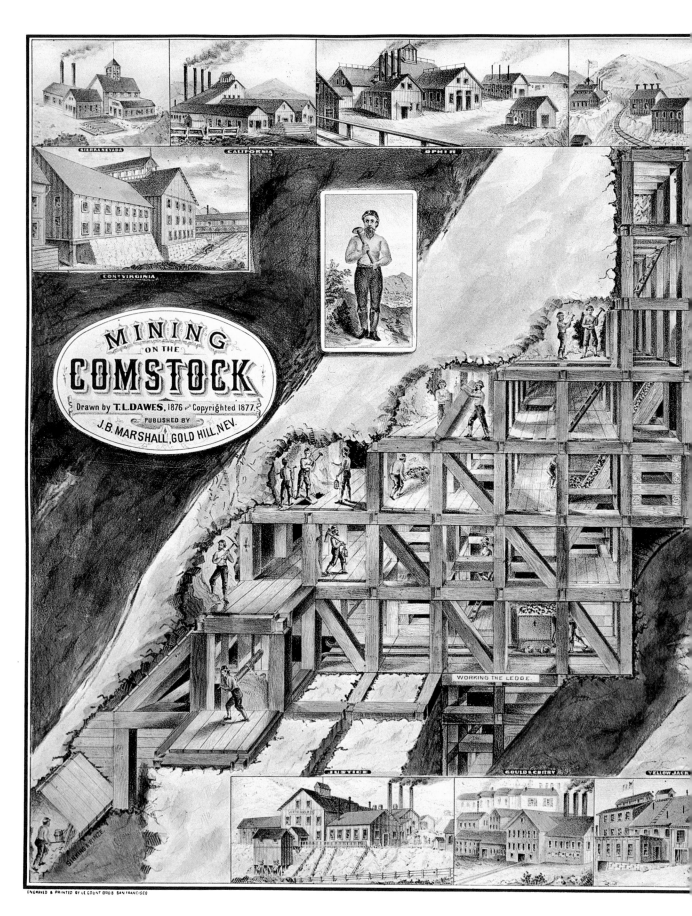

MINING
ON THE
COMSTOCK

Drawn by T.L.DAWES, 1876 AND Copyrighted 1877.
PUBLISHED BY
J.B. MARSHALL, GOLD HILL, NEV.

WORKING THE LEDGE.

houses with one-inch plank partitions, froze in outdoor privies and collapsed of heat exhaustion 2,000 feet underground. Their occupational diseases were pneumonia and rheumatism. Had it not been for their unions and their high wages, which mineowners repeatedly but unsuccessfully tried to reduce, miners of Washoe would have had a bad time.

As it was, they were ground down by boredom and the inhospitality of their environment, and they regularly sought solace in drink. In 1880, which was considered a dry year in terms of thirsts quenched, some 200,000 gallons of alcoholic beverages were brought in by railroad. "Heavy drinking was the curse of the Comstock," said a critic, who noted that some miners put away a quart of whiskey a day.

Predictably, Virginia City had an arrest rate considerably higher than that in an ordinary city. In one year, 1863, the number of arrests was equal to one third of the population, though many of the offenders were not miners but transient ruffians and the number arrested included many repeaters. The charges, in descending order of frequency, were: drunk and disorderly, disturbing the peace, fighting, and sleeping on the sidewalk—all of which reflected booze and tedium more than viciousness.

Perhaps the most impressive objects carried into Virginia City by railroad were pieces of machinery. The stationary steam engine reached, on the Comstock, levels of size and strength that are hard to believe. One engine, used for pumping, had two huge cylinders, one of which was 100 inches in diameter, with a stroke of eight feet and a weight of 43 tons. ◉

COMPLEX INNARDS OF A COMSTOCK MINE

The unique engineering required by the lode is detailed in this 1876 lithograph. The large, unstable ore bodies called for a support system known as the square set—short timber joined into a boxlike frame—within which men could work safely. Ore was sent down chutes or hauled up in buckets for loading into cars in the main lateral tunnel, or drift (*lowest full level shown*). From there, it went to a shaft (*lower right*), for reloading into a giraffe, a car designed to travel at an angle (*inset, showing tools*). The giraffe was then drawn up to the main vertical shaft (*far right*), where the ore was then hoisted to the surface. Because of the heat, men had to cool off with ice in a special chamber (*inset, center right*). For ventilation and communication the miners sank winzes (*bottom left*) along the vein from one level to another.

Glimpses of a world "hundreds of feet below daylight"

The nether world of the Comstock Lode never lacked for chroniclers. Journalists by the score descended into the sweltering depths and returned with vivid descriptions of the mines and of the men laboring to extract their treasure of silver and gold. But words could convey only so much. It remained for a young photographer named Timothy O'Sullivan to make the Comstock come alive in 1867 with a collection of eerily dramatic pictures made in a black pit that was, he marveled, "hundreds of feet below daylight."

At 28, O'Sullivan had already become a seasoned veteran of his craft. He had served with the famed photographer Mathew Brady under fire during the Civil War, narrowly escaping death at Bull Run when the cumbersome box camera he was manning was shot away by Confederate artil-

lery. The Comstock, too, held perils —chief among them the ever-present danger of cave-ins. But it did offer one most unusual boon—at least for a photographer. Down in the depths waited a ready-made darkroom as big as the inside of a mountain.

To make pictures in those days, a photographer first coated a glass plate with sensitizing chemicals, then developed the plate immediately after exposure while it was still wet. Both steps had to be performed largely by feel in a darkroom or a lightless tent. But though O'Sullivan had a ready-made darkroom, he also had a problem. The exposure demanded light, and lots of it. And the candles that enabled the miners to do their work were hopelessly inadequate for the needs of photography.

But the young chronicler had a solution, suggested by a fellow photog-

rapher, Charles Waldack, who had succeeded in taking views inside Kentucky's Mammoth Cave the year before. As he prepared himself for the descent into the mine, O'Sullivan added some extra items to his equipment: shiny tin reflectors, a flint-and-steel spark striker and a supply of explosively flammable magnesium.

Deep in the gloomy blackness, he surrounded his subjects with reflectors, prepared his plates, loaded his camera, opened his shutter—and rapped a load of magnesium with his spark striker. The sudden blaze of light alarmed the mineworkers; they dreaded fire in the timbers. But the danger was well worth it. O'Sullivan's pictures reflect the Hades-like atmosphere of the lodes—their somber moods, their steamy heat, their oppressive, buried loneliness—as only the miners had previously known them.

Before going underground Timothy O'Sullivan pictured miners who were waiting at the hoisting shaft of the Comstock's Savage mine.

Jumbled timbers mark a Comstock cave-in. The foot belongs to a clean-up crewman.

Hacking with his pick, a miner stands perilously close to an overhang of crumbly ore.

His work illuminated by a single stub of candle, a miner uses a timber for a seat as he attacks the face of a tunnel in the Savage mine. The crude wheelbarrow at his left was used to transport ore to the hoist.

The cylinders were linked to a 40-foot cast-iron flywheel of 110 tons that helped drive a colossal pump rod, made of 16-inch timbers strapped together with iron plates that went 2,500 feet down a mine shaft. Rising and falling at a leisurely 10 strokes a minute, the half-mile-long rod could lift two million gallons in 24 hours. Smaller pump parts were shipped by rail, but the flywheel could not pass through tunnels. The solution was to build a foundry in San Francisco, disassemble it, freight it to Virginia City, reassemble it and cast the flywheel there.

The idea of machinery on so grand a scale, of Deidesheimer's square sets, of shafts that reached depths of 3,000 feet, and of curiosities to be seen so far underground, tempted many a visitor to go down and have a look. For armchair adventurers, the best guide was William Wright, a writer who got to the Comstock in 1859 and stayed nearly 40 years. Wright was better known by his pen name, Dan De Quille, which he used as reporter and editor for Virginia City's leading newspaper, the *Territorial Enterprise,* and as author of a book on the Comstock called *The Big Bonanza.* He knew nearly everyone in town, and was well liked. A tall, cadaverous man who wore a cloak that made him look like a bishop, Dan went everywhere, saw everything and wrote of it clearly and well.

"The popular idea of a silver-mine among most persons in the Atlantic states," Dan observed, "appears to be that a deep hole in the form of a common well has been sunk somewhere on the side of a mountain, from the bottom of which is dug the silver ore. They suppose it to be hoisted to the surface in buckets, by means of an ordinary windlass or some such rude contrivance. What really is seen at the mine shaft or entrance to one of the leading mines on the Comstock lode is very different."

From the outside, Dan pointed out, a silver mine resembled a factory with a large building covering the shaft head and hoisting works, and with several wings in which carpenters, blacksmiths and machinists worked. The large building was one enormous room, "floored as handsomely as though it were a church," with a lofty ceiling 40 feet above. "Almost the first object that attracts our attention upon entering the place is the mouth of the main shaft. We see rushing up through several square openings in the floor great volumes of steam. This steam appears to be hissing hot, and rushes almost to the roof of the building. We are surprised to see men coolly ascending and descending the very heart of these columns of steam."

The mouth of the shaft proved to be "an opening in the floor about five feet in width and twenty feet in length, divided into four lesser openings or 'compartments' by partitions that run from the top to the bottom of the shaft." Three of the compartments contained cages for hoisting or lowering men, ore and supplies. The fourth was for pumping: it contained an iron pipe a foot or more in diameter, and beside it the ponderous pump rod that forced water up from the depths.

"The hoisting engines are at the end of the building opposite that occupied by the shaft and 50 or 60 feet away," Dan continued. "Here we find the alert and keen-eyed engineers constantly at their post by their engines." Each engineer kept his attention fixed on a large dial that showed the location of his cage in the shaft; at the same time each listened for signals that were struck on a bell beside him. The bell, operated by a pull rope, was the engineer's only means of communication with the men below. "The lives of the miners are in the engineer's hands every minute of the day and night," wrote Dan De Quille. "To turn his head to nod to an acquaintance might cost a dozen lives. The man who is trusted at one of these engines can be trusted anywhere, and to say that Mr. Jones is an engineer at this or that mine is to say that Jones is a man much above the average."

Near the engines were enormous spools, perhaps 15 feet in length, on which the hoisting cables were wound and unwound. The cables, made of braided steel wire, were flat, like tapes, five or six inches wide and three quarters of an inch thick. Their inventor was a genius named A. S. Hallidie, beloved by San Franciscans as the creator of their cable-car system in 1873.

The cages were simply iron frames with solid floors and open sides. They moved at such speeds that a man who thrust an elbow, foot or head beyond the cage lost it as it hit a timber in an explosion of flesh and bone. "As we dart along down the shaft we get a glimpse of what appears to be a room of considerable size, see a few men standing about with candles or lanterns in their hands, hear voices, and probably the clank of machinery. An instant after, all is again smooth sailing

For three decades, as an editor of the Virginia City *Territorial Enterprise,* William Wright recorded the fortunes of the Comstock. A believer in the prudence of an alias, he by-lined stories Dan De Quille.

and we see only the upward-fleeing sides of the shaft."

The cage stopped at the 1,500-foot level where there was a main drift that ran north-south along the line of the lode. At intervals along the drift there were crosscuts, running east-west, and winzes—short shafts —that ran up or down to connect with other levels. The main drift and the crosscuts had narrow-gauge tracks on their floors for the cars that took out the ore. Adjacent mines cooperated with each other, in ventilation and sometimes in lifting rock, and their drifts were connected. Some major tunnel thoroughfares under Mount Davidson stretched for three miles.

Candles and lamps burned everywhere, 24 hours a day. The temperature at 1,500 feet was over 100°. In that heat miners wore breechcloths or long underwear bottoms cut off at mid-thigh. "All are drenched with perspiration, and their bodies glisten in the light of the candles as though they had just come up through the waters of some subterranean lake. Superb muscular forms are seen on all sides and in all attitudes, gleaming white as marble. We everywhere see men who would delight the eye of the sculptor." To keep dirt out of their hair, miners wore a head covering: narrow-brimmed hats, or felt skullcaps cut from ordinary hats. To protect their feet from sharp quartz they wore shoes. That was the extent of their safety equipment.

"At the depth of from 1,500 to 2,000 feet the rock is so hot that it is painful to the naked hand. In many places, from crevices in the rock or from holes drilled into it, streams of hot water gush out. In these places the thermometer often shows a temperature of from 120° to 130°. It is as hot as in the hottest Turkish bath." To make it possible for men to work, great blowers on the surface forced cool air into the mine through pipes two feet in diameter. Despite this, the men could work only half an hour at a time and then had to rest for half an hour. While resting, each man drank several pints of water and chewed ice, placed in barrels near the tubes of the blowers. In one mine, the Consolidated Virginia, about a ton and a half of ice a day was lowered to the men. When their rest period was over, the men walked back to their jobs carrying fast-melting lumps of ice in both hands.

The Consolidated Virginia became Dan's favorite mine, and he grew particularly fond of the 1,500-foot level because it was there that the men cut into the Big

Bonanza, which would yield more than $100 million before it was exhausted.

Glancing around him in rapture, Dan described the scene. "For a distance of 75 feet on each side of us all is ore, while we may gaze upward to nearly that height, to where the twinkling light of candles shows us miners delving up into the same great mass of wealth. On all sides of the pyramidal scaffold of timbers to its very apex, where the candles twinkle like stars in the heavens, we see the miners cutting their way into the precious ore—battering it with sledge hammers and cutting it to pieces with their picks as though it were but common sandstone. Silver ore is not, as many may suppose, a bright and glittering mass. In color the ore runs from a bluish-gray to a deep black. The sulphuret ore is quite black and has but a slight metallic luster, while what is called chloride ore is a kind of steel-gray with, in places, a pale green tinge.

"Throughout the mass of the ore in very many places, however, the walls of the silver caverns glitter as though studded with diamonds. But it is not silver

A CITY BUILT ON SILVER

With seemingly endless wealth under its streets, Virginia City quickly grew into a metropolis of 20,000 in the early 1860s—and was quickly and more lavishly rebuilt after a fire in 1875. This view, looking down Six-Mile Canyon, shows the Comstock capital in 1878. As in any mining town, most of its tradesmen were bartenders: the community boasted 100 saloons, three undertakers and four churches.

Four of the city's proudest edifices are shown in the insets. At upper left stands the International Hotel, which was rebuilt in 1876 and included such ap-

Young America Fire Company

pointments as huge walnut beds, marble-topped dressers, Nottingham-lace curtains on gilt rods and hydraulic elevators. At upper right is Piper's Opera House, opened in 1867 and lost twice to fire; it drew such performers as humorist Artemus Ward, actor Edwin Booth and the renowned striptease artist, Adah Isaacs Menken. At bottom right, at the end of C Street, is the Fourth Ward School, whose bell first rang in 1876. At lower left, the Young America Fire Company assembles in full-dress parade in 1862 before its brand-new pumping engine.

Fourth Ward School

that glitters. It is the iron and copper pyrites that are everywhere mingled with the ore, which in many places are found in the form of regular and beautiful crystals that send out from their facets flashes of light that almost rival the fire and splendor of precious stones. There are also found in the mass of the ore great nests of transparent and beautiful quartz crystals that are almost as brilliant as diamonds."

However beautiful the candle-lit ore may have appeared to Dan De Quille and other reporters, it was extracted at great human cost. The number of men who worked underground on the Comstock was probably never much over 2,000 in the boom years of the 1870s. Possibly 10,000 men all told worked on the Comstock at one time or another. Yet at least 300 of them were killed and 600 maimed or crippled. There are no accurate records because no one bothered to keep them. There were periods when a man was killed every week, and another grievously injured every day.

Falls were the most common cause of death. At the end of their shifts, as they came up swiftly from the caldrons into cool air, miners sometimes became faint and fell. Their bodies ricocheted down the shaft, being torn apart by repeated impacts on the timbers, until a rain of fragments reached bottom and fell into the sump, a pit full of hot water. Small grappling hooks were kept on hand to recover the pieces, which were rolled in canvas or put into wooden candleboxes to be taken above.

Men sometimes fell or slid into the sumps from short distances, but still could not survive. The *Territorial Enterprise* reported in April 1877 that a man named John Exley, working at the 1,900-foot level of a mine, slipped partway into a sump where the water temperature was close to 160°. Although he sank only to his hips and was immediately pulled out, the skin fell off his legs and he soon died.

Despite the strength of Deidesheimer's square sets, a good many men were killed in cave-ins. Most mine superintendents followed his recommendations carefully, but a few did not, failing to use proper timbering or to keep the sets wedged tightly up against the ore. And the complicated, unstable character of the Comstock made it dangerous. For example, the lode contained thick sheets of clay that swelled on exposure to air, exerting tremendous pressure on the timbers. In places where tunnels were cut into it, the clay con-

stantly moved to fill them up, and it was necessary to keep crews at work hauling away the encroaching mass.

Sometimes there were warnings of cave-ins. The timbers groaned and creaked, perhaps for days beforehand. Then, too, the men underground could get a fair notion of what was about to happen by watching the rats. The mines were inhabited by thousands of rats that lived on the scraps from the lunch pails and were quite tame, as rats go; when a miner sat down to eat he could see their bright eyes gleaming in the dusk at the edge of his candlelight. Apparently the rats were finely attuned to faint movements in the rock. When they sensed that a section of the mine was settling, they became uneasy and were seen skittering about, looking for safety. But neither the squeak of rats nor the creak of timbers prevented men from being buried alive.

Dozens of miners lost their lives in fires. Candles left carelessly lighted or a spark from an underground blacksmith's forge could ignite mine timbers or touch off an unsuspected pocket of flammable gas. In a fire that took place in 1869 on the 700-to-1,000-foot levels of three interconnected mines, 45 men were killed. Other common causes of death were the premature explosion of blasting charges and the falling of timber, tools and other objects down the shafts. In 1880, in a mine called the New Yellow Jacket, a car loaded with steel drills was being hoisted to the surface. Near the top it caught on an obstruction—probably one of the drills struck a timber—and spilled its contents into the adjoining compartment of the shaft. The drills fell half a mile and struck eight men in a car on the 2,800-foot level, killing five instantly and injuring the others.

In another accident, ludicrous to strangers but tragic to those who knew them, two poor souls were killed by a dog. The animal tried to jump across a shaft, missed, fell 300 feet and landed on them. Year after year, the toll was steady and frightful.

The death of the Comstock itself was slow. Flooding had been a problem almost from the beginning—water had been encountered at depths of only 50 feet —and as the shafts sank deeper, more water, at increasing temperatures, poured into the diggings. Larger and larger pumps were brought in; and in many mines, drainage tunnels were dug eastward to allow the water to run out and down the flank of Mount Davidson.

In 1864 the Gold Hill *News* suggested that a major tunnel, which would be capable of draining all of the mines, be dug to solve the problem permanently. The idea was taken up by Adolph Sutro, a brilliant and aggressive entrepreneur who for a while had owned an ore-reducing mill on the Carson River. Sutro persuaded the Nevada legislature to give him a franchise to drive a tunnel from the lode to the river, a distance of nearly four miles. The tunnel would not only provide sufficient drainage and ventilation, Sutro believed, but it would also serve as a convenient passageway to haul ore to the mills.

He proposed to finance the tunnel with borrowed money and stock, and to pay for it by charging the mineowners a royalty of two dollars a ton on all ore produced. However, Sutro's abrasive personality so antagonized the owners that they blocked his efforts to raise funds. The project was stalled until 1871 when Sutro, having obtained money in England, finally began to dig in earnest. The task proved so difficult that it took seven years for the 20,489-foot tunnel to reach the Comstock, where on July 8, 1878, it connected with the mine works 1,650 feet below the surface.

If the Sutro Tunnel had been completed a few years earlier, it would doubtless have saved millions of dollars in pumping expenses and might have been profitable. As it was, the tunnel reached the lode after all the major ore bodies above 1,650 feet had been removed; the mine shafts were now deeper than the tunnel, in a few cases 1,500 feet deeper. The tunnel did do some good by making pumping easier. When its north and south laterals were completed, the pumps no longer had to lift water all the way to the surface but could discharge it 1,650 feet below into the Sutro, which drained off as much as four million gallons a day.

Sutro's creation proved to be of little value in ventilation; by the time air passed through the long tunnel, it was as hot as that surrounding the miners. Moreover, no ore was ever removed through it; nor did Sutro discover any ore as he dug toward the lode. But he lost nothing in the venture. Before the value of the tunnel company's stock collapsed, he quietly sold his holding and retired, a millionaire, to San Francisco.

After 1878 the problem of water remained, but with a different emphasis. The quantity could perhaps

For the long drop to the Comstock catacombs, these miners shed nearly everything but their pants. The deeper they went the hotter it got.

A hand-pumped inspection car, fitted out like a surrey for the comfort of stockholders, emerges from the mouth of the Sutro Tunnel. The instrument on the block at left is a theodolite to check the cut's course.

have been managed indefinitely by pumps and the tunnel combined, but the temperature was another matter. Beneath the Comstock there still burned the subterranean fires whose upwelling had formed the lode 60 million years earlier. As mine shafts went deeper the heat increased by five degrees for every 100 feet. At the 3,000-foot level clouds of steam obscured the drifts. In one mine the daily ice allotment went up to 95 pounds a man. Wooden pick handles became so hot that miners had to use gloves. Working time was reduced to 15 minutes out of each hour, and water heated to almost 160° spurted out of the drill holes. In 1882 the report of the Director of the U.S. Mint said, "It is a serious question whether it is advisable to continue work in those mines. The ledge matter at these great depths is of great width but it contains only seams and stringers of ore, in most cases of very low grade."

On the south end of the lode, where Old Virginny Finney and Henry T. P. Comstock had staked their placer claims 23 years earlier, the pumps were shut down in 1882 and the hot water was allowed to rise to the level of the Sutro Tunnel. Thereafter the miners concentrated on removing marginal ore from the edges of worked-out bonanzas above. On the north end of the lode, where Patrick McLaughlin and Peter O'Riley had been troubled by the strange blue sand, the pumps kept going until 1884. In the center, sealed off by bulkheads, pumping continued for two more years. One shaft was driven to a depth of 3,080 feet and water at 170° was encountered, but nothing worth mining was sighted through the steam. On October 16, 1886, pumping ceased and the last hope of finding a bonanza below the Sutro Tunnel was abandoned.

However, mining continued on the Comstock for decades. New refining processes made it possible to work lower-grade ore, and small pockets of high-grade matter were found in places where the old drifts and crosscuts had missed them. By 1900 production was averaging $500,000 a year, and it appeared that the return might continue at that level far into the future.

The Comstock yielded very large fortunes for half a dozen men, and created perhaps 20 ordinary million-

aires. In addition there were hundreds who emerged, either from the mines or from speculations on the mines in the San Francisco stock market, as wealthy men by 19th Century standards, possessors of bank accounts of $100,000 or so. But the greatest beneficiary of the Comstock was the mining industry itself. As the Comstock mines began to fail, superintendents, foremen, mechanics, carpenters and miners, experienced and highly skilled, went out to open new mines in Idaho, Montana, Colorado, Utah, Arizona, British Columbia and Alaska. Indeed they went far overseas as well, and soon Americans were in charge of mines in South America, China, Japan, Africa and Australia.

Dan De Quille went nowhere. He seemed unable to tear himself away from Virginia City. For years he had bached it as a "Washoe widower," living in hotels or rooming houses and sending money home each month to his wife and three children in Iowa. After the mines fell on hard times, and even after the *Territorial Enterprise* folded in 1893, Dan remained on the Comstock. He did not leave until the late 1890s, when he

departed for the home of a daughter in West Liberty, Iowa. There he died in 1898 at the age of 69.

Dan had cherished some curious fancies about the Comstock. Because the mines were "the tomb of the forests of the Sierras," he liked to imagine that other miners in the vastly distant future might return to Mount Davidson and find another bonanza there: timbers turned to coal by subterranean pressure.

He himself sometimes explored the upper levels of the mines, which had been abandoned and were collapsing even while work was going on far below. "Down in these deserted and dreary old levels, hundreds of feet beneath the surface," he wrote, were "fungi of monstrous growth and most uncouth and uncanny form. They cover the old posts in great moist, dew-distilling masses, and depend from the timbers overhead in broad slimy curtains, or hang down like long squirming serpents or the twisted horns of the ram." His candle flame, almost stifled by the fetid air, flickered and sank low. From the darkness beyond its light he heard no sound but the steady drip of water.

The grim, grand works high on Mount Davidson

If people ever thought of beauty in connection with the Comstock Lode, these musings were apt to focus on the richness of bullion and what it would buy. Finding anything beautiful about the source of it all, scabrous Mount Davidson and the mines at its flanks, was another matter. It remained for the West's most renowned landscape photographer, Carleton Watkins, to find grandeur in the upper works of the lode.

Watkins' photographs bestowed on these drab structures their rightful dignity of order and purpose; they emphasized two salient facts about the mines. One was their intimacy with the town they supported; in many cases, mine buildings were surrounded by Virginia City's shops and homes. The second was that once the ore was hoisted out of the mountain, gravity became a major tool in refining it. Virtually every step in the process, from the shaft head to a brick of bullion, ran downhill.

A timber-railed tramway swoops down High Cedar Hill to feed ore to Sacramento Mill.

Zigzagging trestles for ore cars filigree the face of Mount Davidson above the Trench mine's mill. The barrels perched atop the mill's roof ridges *(center and right)* were filled with water and served as a fire-extinguishing system in case cinders from the smokestacks set fire to the shingle roofs.

83

In the midst of miners' homes, the hoisting works of Hale & Norcross lays a pall of smoke and steam over Virginia City. From the shaft head, the ore was run out on tracks and dumped down a chute into holding bins *(left)*. The railed track curving from the right brought in timbers for the mine.

Ore wagons, drawn by 10-mule teams, stop at loading chutes of the Chollar mine, which by Comstock standards was a moderately good producer, yielding $16 million worth of silver between 1861 and 1876. The four-story building above the storage shed was the Fourth Ward School.

Near a brewery and a truck garden stand the upper works of the C & C shaft. Ore was brought to the surface in the cupolaed hoisting house *(center, right)* next to the four-stack powerhouse. Virginia City had many mountains of waste-rock residue such as the one below the hoisting house.

3 | Machines for the hard-rock miners

Hard-rock mining, which took thousands deep into the earth, was scorned by the stream-side placer miners. But the underground workers agreed with a German miner who described their trade as "a calling of peculiar dignity." The job was dignified by constant danger and bone-wrenching labor. During eight- to 10-hour shifts, the miners toiled in hot, dripping tunnels, liable to maiming or death from cave-ins, noxious fumes and misfired charges of dynamite. And after the ore was hauled to the surface, they had to endure the perils of giant machines used to separate the blends of gold and silver from the rock in which they were embedded.

Miners in hard rock could also boast of being modern experts in an ancient craft. The refining had been pioneered centuries before by the Spaniards in Mexico and Peru. There, the slaves crushed the ore with iron battering rams, and mules ground it into a mud-like slurry by dragging flat boulders around a wet stone pavement called an *arrastra*. The Spanish miners then trod salt, copper sulphate and mercury into the slurry to amalgamate with the gold and silver. Finally they gathered up the amalgam and boiled off the chemicals, freeing the precious metals.

The Western miners' industrialized version of these techniques typically required an installation that cost about $200,000 and utilized some 20 acres. One such mill could produce as much wealth in six hours as a Spanish mining complex could turn out in a month.

Ranks of covered, steam-heated pans stretch the length of the huge Brunswick mill in Nevada in the 1870s. The pans received crushed ore from stamping machines (*barely visible in the dark at right*) and ground it into a fine paste from which the rich metals were chemically extracted.

Of giant powder, Cousin Jack, and "that d--ned old Hearst"

We heard yells coming from the tunnel and saw men rushing toward the mouth. In a few seconds there came the sounds of breaking timbers, then a grinding rumble, followed by a depressing silence and a violent rush of air as if a big blast had gone off. Our carbide lamps went out but we did not need light to know the tunnel had caved."

Frank Crampton, a hard-rock miner who worked in Goldfield, Nevada, around the turn of the century, remembered the sound and the sudden darkness for the rest of his life. Crampton and 19 others were trapped deep in a mountainside, their only exit blocked by 400 feet of cave-in that had obliterated their tunnel. When they got their lamps lighted again, Crampton looked at his companions. "Fear was written on every face, as I knew it must have been on mine, but there was no sign of panic. All of us were in for trouble, and there was not one who wasn't scared stiff."

The air was heavy with smoke from the day's blasting and the smell of water-soaked timber. Water was dripping from the ceiling and coming in steady trickles out of the walls. Crampton knew that it would take days or even weeks for rescuers to dig and blast their way through the collapsed tunnel, timbering it as they went. For a moment he feared the water might rise and drown him, but then took comfort in his knowledge that the tunnel had a gentle slope toward its mouth; probably the water would drain out through the caved section. There would be more than enough to drink, surely, but the constant dripping would keep the miners' clothing damp. Food was the greatest problem. When the lunch buckets were inventoried the supply

---

Set off against the photographer's props and fancy background, a hard-rock miner cuts a splendid figure with his sledge hammer and imposing array of steel drill bits.

was very small; each man brought only enough for his midday meal. And only three men had carbide lamps, which would burn for a few hours. The others had candles, which, lighting one from the dying stub of another, would give light for less than four days.

In the first few hours, recalled Crampton, three of the miners "fashioned crude musical instruments. Two were made from powder cases to which necks had been attached, and each had four strings made from tightly drawn rawhide shoe laces. The third was a flute, fashioned from a piece of pipe with a four-inch-long crack. One end was plugged and the other had a very creditable mouthpiece."

On the first day the men sang and listened to the deep bass notes of the powder-case viols and the high-pitched flute. Far off, as the distant concussions traveled through the rock, they could hear the sounds of a rescue party working toward them. But instead of being cheered, the trapped men were depressed. The blasting was heavy, indicating that the tunnel was blocked with huge pieces of rock. Crampton estimated that it would take three weeks to make a breakthrough.

"The third day was less than half over when an extra heavy shot sent a concussion through the workings that put out our lighted candle. Before that no one had given a thought to matches—no one smoked in the mine at any time; we chewed plug or used snuff. The only time matches were used was to light lamps or candles, and that had not been necessary except immediately after the cave-in." Now, when they searched their pockets, they found all their matches water-soaked and useless.

In the absolute darkness, conversation and music stopped. Soon the men became aware of a new, maddening sound. "Whether any of us had noticed the ticking of watches, all 20 of them, I don't know. But with darkness, sounds from our watches were as loud as a

93

boiler factory in which a drum corps was practicing."

After a few hours Crampton crawled from man to man, collecting the watches. All seemed glad to give them up, and there was general relief when he dropped them into a deep, water-filled hole in the floor.

Crampton's sodden clothing constantly rubbed his skin. "It was softening otherwise hard flesh all over my body, and the flesh became so loosened that it would slip and crawl over the meat underneath whenever I moved. My teeth were chattering almost constantly, as if I had malaria. My jaws were so tired and sore that I tied a bandanna around my head and chin to keep them from moving and to stop the agony."

Their food long since gone, the men lay silently in the dark, waiting. They lost track of time. "I have no way of knowing when it happened," Crampton wrote, "but it must have been well along on the ninth day. There was a piercing shriek, then unintelligible words. After a few moments of silence a dozen voices rose to question what was going on. Then another shriek, and sounds of a man running, the sound of a body falling or hitting something, and another shriek followed by deep moans. More sounds of running, more shrieks, more sounds of a body falling or hitting something. All the time the sounds grew dimmer and dimmer; finally, a scream that pierced the workings as would the whistle of a freight engine. Then silence.

"I waited for more sounds to follow, hoping they might, and yet hoping that none would come. Nothing came. The silence was deadly. There was not one of us who did not know what had happened. One of the stiffs had come to the limit of strain, and his nerves and mind had broken. He had gotten up, started to run, and beaten himself when falling to the floor or against driftwalls, time after time. I put the bandanna over my mouth and tied it behind my head to keep from making audible sounds. I wanted to moan, or to cry; it didn't matter which."

On the 14th day the miners' misery came to a blessed end. Crampton felt a sudden rush of ice-cold air and heard a man shouting, "Cover up your eyes, here we come with lights!" Later, in a daze on his way to the hospital, he asked how many men had been saved. "All but one," was the answer.

Crampton's experience was by no means unusual in the hard-rock mines of the West. It was simply an oc-

cupational hazard to be expected by men laboring underground in Virginia City, Goldfield, Rawhide and Tonopah, Nevada; in Coeur d'Alene and Moscow, Idaho; Butte and Helena, Montana; South Pass, Wyoming and Ophir, Utah; Central City, Leadville, Cripple Creek and Telluride, Colorado; Chloride and Tombstone, Arizona; Hermosa and Orogrande, New Mexico, and in thousands of other mines scattered across the mountain and high-plateau country. They all ran the same risks, and the toll was fearful.

No one, certainly not the mining companies, kept track of all the casualties. For decades, a miner's skin was regarded by his bosses as no one's responsibility but his own. However, informed observers guessed that if a man spent 10 years in the mines, he stood one chance in three of suffering a serious injury, and one in eight of getting himself killed. In one terrible explosion and cave-in at a Utah mine in 1903, 35 miners lost their lives. All told, possibly 7,500 men died digging out the silver and gold on the Western mining frontier — with another 20,000 maimed. Not even logging exacted such a frightful human price for its treasure.

Yet the only way to get at the precious metal deep underground was to descend into the dangerous depths and dig it out, and everywhere — except on the Comstock, which presented a unique set of problems — the miners' methods were substantially the same.

Like all sensible men, Western hard-rock miners had an invariable object: do no more work than necessary. They tried always to arrange matters so that heavy ore and waste rock had to be lifted as little as possible. If they were working on a lode that dipped down at an angle, they tunneled into the hill or mountainside to strike the bottom of the ore body. Then they dug upward, letting the rock fall into chutes from which it could be fed by gravity into hopper cars and taken easily out of the tunnel. Mules, more intelligent and sure-footed than horses, were commonly used to pull the cars.

If the miners could not reach the lode by tunnel they sank a shaft to a point below the proposed mining area, dug laterally until they got under it, and again worked their way upward. In this case they faced the added labor and expense of hauling the cars or ore buckets up the main shaft, by hand-turned windlasses in the smallest mines and by huge steam-driven hoists in the larg-

A miner demonstrates the tiring task of hammering a drill into rock to prepare a blasting hole; behind him two men team up on the chore.

Cripple Creek miners of the 1890s assemble around a pair of compressed-air drills mounted on an iron column by universal joints. The speedy pneumatic drills put an end to laborious hand-jacking—and in the process they drastically reduced the number of jobs available to men in the mines.

est. Nevertheless, the arrangement was still better than digging down through all of the rubble from a blast, which the miners had to clear away before they could set another charge.

Until about 1875, when machine drills and dynamite came into widespread use, the principal tools of the industry were hand drills and black powder. Blasting holes were cut in rock or ore by "single-jacking," in which a miner held a drill in one hand and swung a four-pound sledge with the other, or by "double-jacking," in which one man held the drill while a companion or two hit it with eight-pound sledges. The drills were constantly turned in the holes so that they would not stick, or "fitcher." A good team of double-jack men could deliver as many as 60 blows a minute and drill

two inches into solid granite in that time, although no one could maintain that pace for long. Ordinarily it took an hour or so to make a 30-inch hole.

Miners' drills — usually called steels — were made of round or octagonal rods sharpened to plain chisel tips that had a slight flare as a further insurance against fitchering. The starter drill, or bull steel, was about a foot long with a 1¼-inch tip. After a hole had been well begun, the bull steel was removed and replaced by a "change" drill that was six inches longer and ⅟₃₂ inch narrower, so that it would follow easily in the hole. The changing was repeated until, at the usual maximum, the last steel was three feet long with a ¾-inch tip. "Down" holes were considerably easier to make than "up" or "flat" (horizontal) ones, but up holes had an advan-

A compressor operator sits with regal aplomb while his assistants, an oiler and a mechanic, hover nearby in this 1890s tableau.

tage: the rock dust fell out of them, whereas it had to be scraped out of the others with long, thin miners' spoons made of copper. After a drill had cut about six inches of hole, it had to be resharpened and tempered by one of the busiest men in the mine, the blacksmith.

When cutting a tunnel the miners ordinarily used a pattern of seven holes and charges. In the center of the face to be blasted, they drilled three holes about two feet apart, arranged in a rough triangle and angled to meet at the apex of a pyramid within the rock. Then they drilled a "reliever" hole at the top of the face, "edger" holes at each side, and a "lifter" at the bottom. With proper timing, the center charges exploded first, making a cavity into which the slightly later blasts from top and sides squeezed the surrounding rock. Finally the lifter blew the rubble out into the tunnel where it could be mucked into hopper cars. In some mines, where the rock was particularly resistant, a 16-charge pattern replaced the usual seven charges.

To provide a flat working surface and to aid in turning the cars, a large sheet of iron plate was laid down at the rock's face and shoved along as the tunnel lengthened. Ordinarily a team of miners drilled the face through a single eight- or 10-hour shift and blasted it just before quitting, leaving the sequence of cleanup, more drilling and more blasting for the next shift.

The blaster in a frontier mine had to be a man of skill and good judgment if he proposed to enjoy a long career. From a keg of powder aboveground or in some secluded corner of the mine, he measured out what he estimated to be the right amount of explosive for the job at hand. Handling it with wooden or copper tools to avoid the danger of a spark that might be produced by iron, he made it up into paper cartridges and fitted them with carefully measured lengths of fuse. When the cartridges were inserted into the holes and tamped with moistened drill cuttings, the fuses dangled out like rattails, as in fact they were called.

The type of fuse employed throughout the West was the Bickford Safety, invented by an Englishman in 1831. It was flexible enough to be wound and carried on a big spool, and consisted of a core of powder surrounded by twisted strands of jute, wrapped with a layer of twine and then wrapped again on the outside with waterproof tape. The fuse burned at a reliably uniform rate and seldom failed; even when it did, it tended to fizzle out rather than burn too fast and set off a premature, lethal explosion.

To ignite the rattails the blaster cut a fuse called a spitter, shorter than all the rest. Then he shouted his traditional warning, "Fire in the hole!", lit the spitter and speedily applied its sparkling end to the rattails in the desired sequence. When the spitter singed his fingers he knew it was time to depart. When the blast came, it advanced the tunnel by about three feet.

In the 1870s, dynamite replaced black powder in the Western mines and was greeted with mixed feelings by the men who used it. The explosive element was nitroglycerin, so hypersensitive that it would detonate if a man so much as spoke harshly in its presence. Pure nitroglycerin, known as blasting oil, had been used experimentally in a few mines, but had proved so risky to handle that few men would touch it. The Swedish scientist Alfred Nobel housebroke it by combining it with inert substances, including chalk, that turned it into a stiff gelatin-like substance. Nobel's dynamite sticks were so docile, relatively, that they could be sliced like bananas, molded into charges shaped for special needs, or tamped down with a certain amount of vigor. To explode, dynamite required a heavy jolt that was usually provided by a small, tubular copper blasting cap containing fulminate of mercury. The cap, which could be set off either by a spark or a light concussion, was inserted in the side of a dynamite stick.

Miners tended to be careless about these caps and often left them lying about where youngsters could get hold of them, as mining chronicler Dan De Quille noted in *The Big Bonanza*:

"The first thing they do is begin probing and scratching in the interior of the little cylinders in order to get out and examine a sample of their contents. It invariably happens that at about the first or second scratch the cap explodes, and the person engaged in prospecting it loses the ends of two fingers and the thumb of the left hand. In Virginia City and Gold Hill about one boy per week, on an average, tries this experiment, and always with the same result."

Miners themselves suffered injuries — or at least embarrassment — because of their casual attitude toward the caps, as De Quille observed. "Miners very frequently carry these caps loose in their pockets, often mixed with their tobacco, and thus get them into their pipes. Several favorite meerschaums have been lost in this way, and the ends of a few noses."

Dynamite could, naturally, be set off by means other than blasting caps. It was unwise to hit it with a sledge, as suggested by this report from a Denver newspaper of 1889: "Albert Scott, a contractor on the Hall Tunnel, met with an accident last Saturday which resulted in his death the following day. He was engaged near the entrance in putting up hooks in the roof of the tunnel to hold a piece of equipment. Holes had been drilled and wooden plugs prepared for them into which the hooks were driven. It is thought that Mr. Scott must have had among his plugs a piece of giant powder about the same size, which exploded when he attempted to drive it into the hole. To one who does not know Mr. Scott's quick and frequently careless way of working, this would appear unreasonable, but it is the only plausible theory of the cause of the accident. The explosion crushed in the upper portion of the skull and knocked both eyes out."

Dynamite, which was about four times as powerful as black powder, had other disadvantages. Its explosion released clouds of nauseating fumes, which in poorly ventilated mines felled men who approached a tunnel face too soon after a shot had been fired. And occasionally one of a group of charges failed to explode, creating a problem for miners in the oncoming shift. The unexploded charge could be located fairly easily —it would be in a mound of rock protruding from the face—but it had to be removed with great care, usually by a miner who dug around it with a pick.

Mining town newspapers contained many brief items such as the one that appeared in a Colorado journal of the 1880s under the headline:

THE FATAL 'MISSED SHOT'
Another name has been added to the long list of victims of that dangerous but unavoidable work of picking out blast holes that miss fire. Frank Benjamin, a highly esteemed citizen of Golden, who has been working in the Shenandoah Valley Tunnel, Red Elephant Mountain, was found dead in the tunnel on Tuesday afternoon. Benjamin was working alone in the breast of the tunnel. Previous to going to work on Monday afternoon he told a fellow miner

An operator in a mining shaft house works the controls of the steam-driven hoist that raises and lowers cages carrying men and ore. Hoistmen like the one shown here at the turn of the century, earned four dollars a day in their highly exacting jobs—a dollar more than colleagues down below.

Mineworkers crowd around the head of a shaft and pack into elevator cages for the ride down to a gold mine at Cripple Creek.

that he had a missed shot to pick out. He was last seen alive about half past three. The deceased leaves a wife and child at Golden. Mrs. Benjamin was engaged in giving a music lesson when the sad news reached her, and her sudden plunge into sorrow is said to have been most pitiable.

Frank Benjamin's body was recovered, but some victims were torn into such small pieces that there was little left to bury, as reported in another Colorado paper on March 18, 1891. "The most disastrous mining accident in the history of Clear Creek county occurred at the Atlantic-Pacific Tunnel, resulting in the killing of Harry Taylor, William Coughlin, John Richards and John Mulholland. The men were employed in driving the tunnel. It is the custom to drill 16 holes and fire them at one shot. To do this requires about 50 lbs. of giant powder. Nine holes had been loaded, and while engaged in tamping the tenth hole, the charge exploded, setting off all the powder that had not been used. The men were torn and mangled beyond recognition. Arms and legs were torn from the bodies of all the men, and pieces of flesh were scattered for a hundred feet along the tunnel. The body of Mr. Taylor was found a hundred feet from the explosion. Another, who was evidently standing over the powder when it exploded, seems to have been thrown against the roof of the tunnel and dropped back a horribly mangled mass of humanity." In such cases it was customary to dust the tunnel with quicklime before work was resumed.

Compressed-air drills, fed air by hose from surface steam engines and using pistons to work the steels back and forth in the drill holes, came into use in the 1870s at about the same time as dynamite. They were greeted with great joy by owners of large mines that employed scores of men. The machines "made hole" at a prodigious rate, enabling the owners to lay off most of their double-jack teams. True, the owners had to hire more blacksmiths because the machines blunted steels after very brief use, but the net result was a loss of jobs and the demotion of proud drillers to lesser and lower-paid work as muckers and car pushers.

Compressed air did bring a few advantages to the miners—exhaust from the drills was refreshing, and air could be taken from the lines to operate blowers and

run small hoisting engines. But the miners soon learned to refer to the machines as widow-makers. As the drills cut into granite, quartz or porphyry, they stirred up clouds of razor-sharp particles of silica dust that lodged in men's lungs and, in time, disabled and killed them. Hundreds of miners died in this manner (the condition, silicosis, was generally called Miner's Consumption) until the mid-1890s, when a water-flushed drill came into use. It had a small, lengthwise hole in its core through which water was forced, wetting down the dust at the drill tip and as a fringe benefit turning it into a wet-grinding compound.

The combination of dynamite and compressed-air drills, while reducing the number of drillers in big mines, actually increased the total labor force. After 1875 it was possible for only four or five miners to handle great quantities of ore at an acceptably low cost per ton. As a result, many lodes of small size, or lodes containing low-grade deposits were opened up and worked throughout the West.

The men who labored in these hard-rock mines were as polyglot a group as could be found anywhere in the country. In fact in some places, native-born Americans were distinctly in the minority. In the Comstock, for example, the census of 1880 showed 2,770 men in the below-ground labor force, but only 770 were Americans. While Americans were expert placer miners from the forty-niners on, they had little experience with techniques of hard-rock mining. Consequently, at the face of the deepest tunnel or in the midst of the most daunting perplexity, there was likely to be a skilled foreigner —often a tough, tireless Englishman from Cornwall.

Cornishmen had been working in the tin mines of southern England for centuries, and their knowledge of deep tunneling and following difficult veins was immense. As the gold and silver mines of the West opened up, Cornishmen emigrated to the United States by the thousands and found jobs immediately. Mine superintendents were so impressed with them that they constantly asked the Cornishmen if they had any relatives at home who might be persuaded to come over. To this the standard reply was, "Yes, I have a Cousin Jack who might." In time all Cornishmen came to be called Cousin Jack and their wives Cousin Jennie.

Cornishmen were very popular on the mining frontier, not only for their skills but for their customs, amiability and flavorsome use of language. They called everyone to whom they spoke by some endearing term —"my son," "my 'andsome," "my beautay"—and they were robust drinkers and melodious singers, filling the saloons with the strains of "The Wreck of the Arethusa" and "Trafalgar's Boy." Ironic wit was their forte. One Cousin Jack, when asked if he could suggest a likely spot to find gold, said, "Well, sorr, where gold is, it is, and where it ain't, there be I." Another, giving testimony in a court case involving the worth of a mine, was asked whether there was an ore vein at the bottom of the shaft. "Not a dom bit," he said, "and smaller as 'ee goes down."

Because so many Cornish names began or ended with certain combinations of letters, a children's verse often chanted in mining towns was: "By Tre-, Pol-, Pen- or -o, / The Cornishman you come to know." This may have mystified strangers, but seemed simple enough to people named Trenoweth, Polglase, Penrose or Fenno. Cornish dialect was also puzzling at times, as when a Cousin Jennie walked into a butcher shop in Nevadaville to buy supplies for her boardinghouse. Spying a stuffed owl perched decoratively on a shelf, she said, "'Ow much for the flat-faced chicken?" The butcher replied, "That's no chicken, that's an owl." At this the lady snapped, "I dunt care 'ow owl he be! Ee'll do for boarders!"

Often the Cornishmen turned their humor against themselves. According to a favorite story of the outlanders, two Cousin Jacks were sitting high on a scaffold in an underground cavern, drilling the face when the plank broke and dropped one of them to the floor. The other hung onto the drill sticking in the hole, but finally had to let go, and when he hit bottom his friend remarked, "Damme, I knew 'ee was slow, but I didn't know it would take 'ee five minutes to fall 25 feet."

Muscular and keen of eye, the Cornishmen were superb at single- and double-jacking, and could swing sledges for long periods. In one hard-rock mine a Cousin Jack was teamed with an American apprentice, who took the easy task of holding and turning the steel while the Cornishman pounded it with the heavy sledge. It was the practice for double-jackers to take turns at the sledge, lest one become worn out, but the American seemed unaware of this. After a half hour of furious work the exhausted Cousin Jack put down his sledge and remarked, "Thee've good wind for turning, my 'andsome."

Drilling contests, frequently won by Cornishmen, were the main events in Fourth of July celebrations in many mining towns, even after mechanical drills came into use. The stone ordinarily used was Gunnison granite from Colorado, cut into a block at least six feet thick with upper surface dressed flat. While a timekeeper stood by, the miners pounded away at their steels for exactly 15 minutes. The world's record for the "straightaway" (two men, but no change of position) was set in Bisbee, Arizona, on July 4, 1903 by a Cornishman named Sell Tarr, who drilled 28 5/8 inches. In a double-jacking event that same day in El Paso, Texas, with both men taking turns on the sledge,

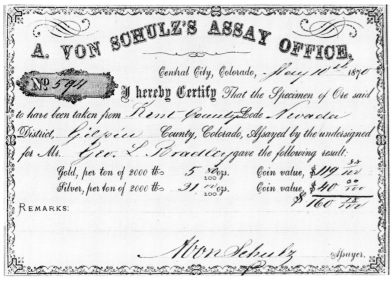

A mineral assayer shows off his laboratory in Blackhawk, Colorado. Assayers, working for a mining firm or in business for themselves, performed complex tests on ore samples to determine how much gold and silver they would yield. Delicate scales, like the glass-cased ones on the counter, were used to weigh the original sample as well as the final extract of precious metals. Findings were formalized with a certificate like the one at left, which indicates the claim of George Bradley contained 5.8 ounces of gold and 31 ounces of silver per ton—more than enough to make mining feasible.

At a Labor Day drilling contest at Leadville, in 1900, one team member called a shaker holds onto a steel bit as his partner prepares to slam home a nine-pound sledge. The best of the drillers, competing for purses as high as $1,000, could bore a two-foot hole in the granite slab by averaging an awesome 76 strokes a minute in the 15-minute event.

Some common forms of raw treasure that were coveted by miners were black-coated silver ore, gold and fool's gold (pyrite), and glittering bluish-gray galena, which held both gold and silver.

SILVER ORE

PYRITE AND GOLD
IN A QUARTZ VEIN

GALENA

the team of Chamberlain and Make sank a hole 42½ inches deep—another record.

Such trials could bring disaster as well as glory. In Nevada a miner's wife, Mrs. Hugh Brown, left this account of an accident-marred contest:

"A drilling contest has everything: technique, beauty, endurance, speed and danger. If the hammer descends a fraction of an inch out of line on the tiny head of the drill, a man's hand may be crushed.

"During my life in Tonopah I saw a man's hand struck. Suddenly the hammer poised in midair. The crowd groaned, knowing what had happened. After an instant flinch, the man crouched over the drill looked up at his towering partner and yelled, 'Come down, you!' Down came the hammer. The men cheered and the women cried. The hand on the drill began to turn

red, but still it held on to the drill. When the injured man's turn came to rise and hold the hammer, the blood crept down his arm until it looked as though it had been thrust into a pot of red paint. The blood ran into the hole and mixed with water from the hose used to flush out the drill cuttings. Every time the hammer descended, the red fluid sloshed up and spattered nearby onlookers. The man sagged lower after every blow, but he never gave up until the timer's hand signalled fifteen minutes. Then he fell over in a dead faint. The platform looked like a slaughtering block."

The great hero in many Cornish households was not a miner but a prize fighter, Bob Fitzsimmons. Although he weighed only about 165 pounds and looked —according to a contemporary—"like a cannon-ball on a pair of pipe-stems," the freckle-faced, grinning Fitzsimmons won world championships in three classes, including heavyweight. Known to thousands of Cousin Jacks and Jennies as "The Cornishman," Fitzsimmons scored his greatest triumph on St. Patrick's Day in 1897 in Carson City, Nevada, when he took the heavyweight title by knocking out the Irishman Jim Corbett. That night there was scarcely a sober hard-rock miner west of the Mississippi.

In addition to the talented Cornishmen, the Western mines attracted Tyroleans from Austria and many unskilled workers from Ireland, Italy, Germany and Serbo-Croatia. The latter, known as "Bohunks," produced the writer Lazar Jurich, whose ballad "Underground in America" lamented the miner's lot:

> For the mine is a tragic house,
> It is the worst of prisons—
> In bitter stone excavated,
> In barren depths located—
> Where there is no free breath,
> With a machine they give you air.
> By you always burns a lamp,
> And your body struggles with the stone.
> Hands work, never do they stop,
> And your chest sorrowfully heaves,
> For it is full of poisoned smoke
> From gelatin's powder white.
> We, miners, sons of sorrowing mothers,
> Look like men from the wastelands.
> In our faces is no blood
> As there is in other youth.
> Many poor souls their dark days shorten,
> Many poor souls with their heads do pay.
> There is no priest or holy man
> To chant the final rites.

At the very bottom of the labor scale were the Chinese, who poured into California in the 1850s and migrated throughout the West, turning up sooner or later at the scene of every new gold strike. Frugal and immensely industrious, they were willing to work for wages so low as to undercut the pay of Americans and Europeans. This did little to make them welcome, and they suffered also from race discrimination, with the result that they soon found themselves effectively prohibited, by threat or by force, from holding decent jobs in any hard-rock mine. Even in placers they were not

permitted to stake good claims, but were obliged to re-work old diggings from which most of the gold had already been taken. But the Chinese often managed to extract good sums from rocky, seemingly hopeless locations. Their diggings were invariably marked by neatly stacked boulders, each of them carefully hand washed to remove all the clay and particles of gold that might have adhered to it.

In many mining communities the Chinese became cooks and laundrymen. However, they were not secure even in these low-paying jobs. In the 1860s in Helena, Montana, a "Committee of Ladies," consisting mostly of irate miners' wives, demanded that the Chinese immediately "suspend the washing or laundry business," issuing an ultimatum that was put into verse by the local newspaper editor:

> Chinamen, Chinamen, beware of the day
> When the women shall meet thee in battle array.
> Ye hopeless professors of salsoda and soap,
> Beware of the fates that await ye;
> No hangman's committee with ladder and rope,
> But the ladies are coming to HATE ye.
> Ye almond-eyed, leather-faced murthering
>   heathens!
> Ye opium and musk-stinking varments,
> We will not object to your livin' and breathin',
> But beware of the washing of garments!

In Virginia City, Montana, another editor reported the arrival of a stagecoach full of Chinese looking for work and invented names for them: Whang, Hong, Lo-Glung, Ho-Fie and So-Sli, while still another suggested that Chinese should not be killed "unless they deserve it, but when they do—why, kill 'em lots."

For sheer malevolence and vileness, the Tombstone *Epitaph* in 1882 set a standard that was hard to match. "The Chinese are the least desired immigrants who have ever sought the United States," wrote the editor. "The most we can do is to insist that he is a heathen, a devourer of soup made of the fragrant juice of the rat, filthy, disagreeable, and undesirable generally, an encumbrance that we do not know how to get rid of, but whose tribe we have determined shall not increase in this part of the world."

Actually, the editor knew full well how to "get rid" of the Chinese, who were lynched and persecuted with

PROCESSING ORE FROM THE TOP DOWN

While Western mining in the 1890s still relied on the backs of strong men for the digging, the chore of processing ore had reached highly sophisticated levels. Here at a gold and silver mill in Eureka, Utah, coal-burning boilers (*directly below*) produced steam that powered this plant to a daily milled output of 170 tons. At the top level of the complex, crushers broke rough ore into fist-sized chunks. A step below, it was smashed in water beneath half-ton stamps, and this resultant mixture was passed through fine screens. Then it dropped onto vanners—oscillating six-foot-wide belts that winnowed out lead and some silver. The remainder, containing all the gold and most of the silver, went into amalgamating pans to be cooked eight hours with mercury and chemicals, then transferred to settling tanks —out of which the tailings were flushed.

BOILERS

With proprietary pride, laborers show off giant pulleys and cams in an Arizona stamp mill circa 1895. The slamming of metal on metal created a din that made "the roar of Niagara sound faint as a murmur."

general impunity all along the mining frontier. In Virginia City, Nevada, about 1,500 Chinese working on the nearby Virginia and Truckee Railroad in 1869 were chased into the hills and held at bay by miners until the railroad promoters agreed to keep the Orientals out of the city and away from the mines. And in another incident in 1885, at Rock Springs, Wyoming, 28 Chinese were killed in a race riot. To be sure, such major incidents were rare, but the Chinese were harassed and killed in smaller numbers in dozens of mining towns throughout the West.

In the raw world of the miner, some of the worst violence pitted the workers against a far more formidable adversary than the hapless Chinese—their bosses, the mineowners. Throughout the 1880s and 1890s, the union movement was gathering force in the United States and testing itself in confrontations with large corporations. The labor magazine *Outlook* reported that

in 33 months at the turn of the century, 198 union men—including many miners—had been killed and 1,966 wounded in battles with corporate guards, state militia or federal troops. Labor strife caused about half as many deaths as the Spanish-American War, and a slightly higher number of wounded.

Labor troubles in the mines followed a depressingly uniform pattern all across the West. Underground workers, who were paid about three dollars a day, would strike to increase the wage to $3.50 in good times, or to prevent a cut to $2.50 in bad. Mineowners would then import scab labor, protected by armed guards, to break the strike. A battle would ensue, and the governor would call out troops, generally to suppress the strikers. Miners' unions were widely regarded, by government officials as well as by mineowners, as dangerous groups of foreign anarchists.

In 1892, when the mineowners in the Coeur d'Alene district of Idaho decreed a 50-cent-a-day wage

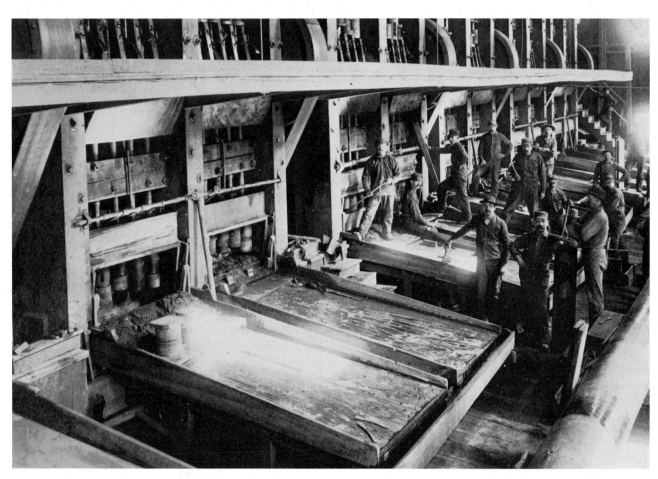

cut—a sum that represented the difference between scraping by and real hardship—the miners angrily went out on strike. In response, the owners brought in a trainload of laborers from as far away as Michigan, including many immigrants from Austria, Poland and Sweden who could speak no English and were unaware of their role as strikebreakers. While they—along with numerous hired guards—were holed up in a mill, the strikers loaded an ore car with 100 pounds of dynamite, lit the fuses, and sent it rolling downhill toward them. One man was killed, 20 were seriously injured, and a large part of the mill was demolished.

This victory for the miners, if indeed it was one, was short-lived. In answer to the requests of mineowners, the governor of Idaho called up six companies of the National Guard; meanwhile, President Benjamin Harrison sent 20 companies of United States infantry. The 600 members of the miners' union, along with some sympathetic merchants, lawyers, saloonkeepers and even a justice of the peace, were rounded up and imprisoned in a stockade, without trial or hearing, for several weeks. Finally, the miners' union caved in. The wage cut remained.

Another ferocious clash took place at Cripple Creek in 1894, after the mineowners suddenly increased the miners' working hours from eight to nine hours with no increase in pay. Close to 500 miners went on strike, and the owners reacted by shipping in 125 well-armed strikebreakers on two flatcars from Denver. The miners greeted their arrival by dynamiting a mine works beside the railroad tracks: as the strikebreakers alighted from the train, an explosion blew the shaft house 300 feet into the air, and a few seconds later another blast sent the boiler sailing after it. In the bedlam that followed, one man on each side was killed by gunfire, and many more were wounded.

During the post-battle lull, the miners took a good defensive position on a high bluff near Cripple Creek

113

A new kind of treasure from a worn-out mine

By 1895 this copper mine in Butte, Montana, was earning more than many a Western gold mine.

Before going into battle at Richmond in 1862, a Union soldier named Michael Hickey read that "Grant will encircle Lee's forces and crush them like a giant anaconda." Hickey was much taken with the name of the great snake, and 13 years later, while prospecting for silver in Butte, Montana, he staked a 1,500-by-600 foot claim and called it Anaconda.

Beneath Hickey's feet lay a virtual mountain of copper. But Hickey did not know—and if he had, he probably would not have cared. In 1875, most of the miners still set their sights and their hearts on gold and silver; they disdained the plebeian "red metal" as a nuisance.

In six years of working the Anaconda mine, Hickey unearthed a modest amount of silver. Then in 1881 he sold out for $70,000 to Marcus Daly, an Irishman who was backed by a syndicate of big-time mining entrepreneurs.

Late in 1882, Daly's silver miners hit a vein of almost solid copper at 300 feet below ground. More digging revealed the ore body ran to a depth of 600 feet and measured an astounding 100 feet wide at its widest point. Meanwhile, the Anaconda's silver output continued to be disappointing.

With foresight Daly and associates put aside their old mining prejudices and built a copper smelter. By the end of 1884, they were shipping huge quantities of bulk copper east and west for such traditional uses as house roofing, cooking utensils, brass hardware and sheathing for ships. At the same time they found an enormous new market opening up for electric cables, appliances and telephone wires. So it was that by 1892 the Anaconda had become the great single copper producer in the world, with an annual output of 100 million pounds.

and armed themselves for all-out warfare by improvising a catapult that fired beer bottles filled with dynamite. Each man was given a personal supply of deadly missiles—five sticks of dynamite fitted with percussion caps that would set off the explosives on impact. And, as a kind of ultimate weapon, they planted explosives in the mines and connected electric wires so they could blow the whole place sky-high at the push of a button. Despite the certainty of grave losses of life and property, the strikebreakers prepared to attack the entrenched miners. But a bloodbath was prevented by a cool-headed militia general named E. J. Brooks, who placed his large force between the two sides and ordered them to disperse.

The Cripple Creek miners' show of strength and determination caused the mineowners to rescind their speed-up order. But that was a stunning exception—as was the benevolent neutrality of the state militia. Indeed, in a later strike at Cripple Creek, the Colorado state militia arrived under the command of General Sherman Bell, who thunderously announced: "I came to do up this damned anarchistic federation." And so he did. With 1,000 troops, he slapped the entire area under iron-fisted martial law, declaring "I'll take no further orders from the civil authorities." One of the general's aides, when reminded of the Constitution, cried, "To Hell with the Constitution! We are not going by the Constitution!" The militia escorted strikebreakers to and from the mines, and scores of miners and their sympathizers were arrested at bayonet point and imprisoned in a stockade. When union lawyers instituted habeas corpus proceedings, Bell roared, "Habeas corpus, Hell! We'll give 'em post mortems."

Hearing of all the labor trouble, newspaper readers back East could hardly help but wonder what had happened to the romantic notion of the Western miner as a keen-eyed prospector searching the pristine wilderness with pick and pan for his pot of gold or silver. He was still there, of course, finding precious metal, and lots of it. But almost from the start, where hard rock was involved, harvesting gold and silver was the work of heavy industry and giant corporations.

Bringing the ore to the surface was only half the task in hard-rock mining. Once the ore was in hand, the millers took over—and if it required armies of men and millions of dollars to tunnel into the bowels of the earth, it took more armies and many more millions to part the gold and silver from its protecting rock. Indeed, few basic industries in 19th Century America could rival the gold and silver millers in the technology they devised or the plants they raised. Nowhere was this side of mining more impressive than in the Homestake mines in the Black Hills of Dakota Territory. As always, the Homestake started with the prospectors —but it ended with the saga of a visionary tycoon who turned it into the biggest of big businesses.

In the late 1870s, rich placer deposits of gold were found near Deadwood in the Black Hills. As the easily worked ground was taken, prospectors fanned out to search for hard-rock sources of the metal. Among the seekers were the brothers Moses and Fred Manuel, who had learned gold mining in California, Nevada and Idaho. Three miles south of Deadwood they discovered a promising quartz vein, or lead, for which the town of Lead was soon named, and they immediately devised facilities to crush the gold-flecked rock.

Like other capital-poor prospector-miners throughout the West, the Manuels did the job with a water-powered *arrastra*—a primitive machine in which heavy, flat-sided boulders were dragged over chunks of ore. The *arrastra* could pulverize ore sufficiently for the liberated gold dust to be collected by panning, amalgamation with mercury, or both—and there was little labor to be expended. But the device had its drawbacks. It could process only free-milling ore—that type whose gold readily separated from the encasing quartz. The Manuels were fortunate to have found the free-milling variety; even so, their machine could not crush it fine enough to extract more than half the gold.

It hardly seemed to matter. The quartz in the Manuels' claim, which they called the Homestake, was so rich that in a few months they took out $5,000 and were able to obtain title to two other claims on the lead, the Old Abe and the Golden Terra. At that point, they sold out at a handsome profit. The Old Abe went to one developer for $45,000, and the Homestake plus the Golden Terra brought $105,000 from a triumvirate of San Francisco entrepreneurs: Lloyd Tevis, James Haggin and George Hearst.

The Manuels' total of $150,000 was a respectable fortune, but it was pitifully insignificant in light of

Following the death by bombing of 13 nonunion miners at a train depot in June 1904, a member of the Mine Owners' Association delivers an anti-union harangue to workers in Victor, Colorado. Soon after this picture was taken, a shot rang out, starting a free-for-all in which two men died.

what George Hearst was soon to do as head of the partnership's operations. All told, his company was to spend more than $650 million to dig and mill Homestake ore; the industrial colossus he put together would eventually take more than one billion dollars out of the Black Hills.

Hearst, the senior partner of the trio, was 57 and had already become a rich man by 19th Century standards as a result of his operations in the Comstock. Shrewd and ambitious, he launched a brilliantly planned campaign. In 1877, he established himself in the town of Lead and soon bought and bullied his way into control of a score of claims beside or near the original Homestake. Many of the claims were disputed or overlapping. Yet Hearst hired the best lawyers and outmuscled his competitors in court. At one time he faced more than 20 lawsuits simultaneously but, by verdict or settlement, was victorious in almost all. In assembling his company, he needed not only mineral rights but millsites and water rights as well, and no speck of land was too small to escape his attention. He paid as little as $25 for one tiny plot, and bought the houses and lots of 16 men in a millsite area for $375 apiece. However, he was generous when he had no other recourse, and paid $106,775 for a mine he considered necessary to his enterprise.

As he gathered his corporation together, Hearst had his qualms, at times thinking he might be murdered by rivals or by litigants who felt they had been wronged. It was typical of the tycoon that he thought men who opposed him were guilty of "fraud," and he wrote to his partner Haggin that "if we succeed in finding out the fraud and maintain our rights there would be more squealing than ever was heard of before. And it is quite possible that I may get killed, but if I should I can't but lose a few years, and all I ask of you is to see that my wife and child gets all that is due from all sources and that I am not buried in this place."

As it turned out, Hearst's life was never seriously threatened, and by 1879, as he neared the completion of his takeover, the *Black Hills Daily Times* had become philosophical. "What difference does it make whose money puts in such great enterprises — whether it is Geo. Hearst's or the Angel Gabriel's — only so we

Armed against saboteurs with bayonet and pistol, a Colorado militiaman guards a Leadville mine in an 1896 lockout of union miners.

get the big mills and the wealth of the country is thus developed? Then why this insane cry against the Californians who have made us what we are? They are our benefactors, and instead of speaking of Mr. Hearst as that 'd--ned old Hearst,' he should be honored, and every courtesy extended to him."

Whatever the case, the Black Hills country got its "big mills," for Hearst brought in some ore-reducing machinery that made the Manuels' simple *arrastra* seem little more effective than a dining-table peppermill. One machine, known as the Blake jaw-crusher, stood shoulder high and could break down large chunks of ore at the rate of 100 tons a day. It had been developed in the East to produce ballast for ships or broken stone for highways and railroad beds. Its "jaws" were a pair of huge opposing iron plates which were brought crunchingly together by an eccentric gear powered by a steam engine.

Once the ore was reduced to pellet-sized (three inches on a side) chunks in the crusher, it was sent for pulverizing to an even more fearsome machine called a "California stamp mill." This device was the size of a house and cost up to $30,000. As the name indicated, it had been developed on the West Coast in the 1850s and brought to a high degree of sophistication in the Comstock a decade later. In principle it was an adaptation of the pharmacist's mortar and pestle, but on an immense scale. Vertical, iron-headed stamps that weighed as much as 1,000 pounds were connected to overhead camshafts that repeatedly lifted and dropped them into iron batteries, or mortars, containing water and mercury for amalgamation.

Hearst's first mill was a large one, containing 80 stamps. It could process more than 240 tons of ore per day, extracting $2,000 worth of gold. But as the veins of the Homestake were opened, it became necessary to add more and more machines—until by 1900 there were no less than 1,000 stamps. Their rhythmical thudding was sometimes audible for miles downwind as they reduced mountains of ore each day, pulverizing it so thoroughly that as much as 90 per cent of its gold could be recovered. To be sure, a 10 per cent loss was substantial, amounting to several million dollars in the early years of the Homestake's operation, but by the

Rubble marks the site of a concentrator in Coeur d'Alene, Idaho, in 1899 after striking miners exploded 3,000 pounds of dynamite.

turn of the century means were developed to cut the loss almost to zero.

To the Western miner, free-milling gold ore like that of the Homestake was second only to loose placer gold on the list of pleasing items to be found in the hills. But when gold or silver was combined with some base metal, or appeared in one chemical disguise or another, as often happened, miners faced all sorts of trouble. After the ore was pulverized, the chemists stepped in, using arcane formulas and vats of eerily smoking liquids to extract the precious metals. No two techniques were precisely alike because ores varied from district to district, and each required a different treatment that had to be determined by trial and error. The silver-gold ores of the Comstock, for example, were subjected to dozens of bizarre doctorings before the millmen hit upon the right one.

The Comstock millmen were early convinced that the gold and silver in their ore could be extracted with the aid of chemicals, but they had no idea which chemicals might be the most effective. Recalled Dan De Quille: "A more promiscuous collection of strange drugs and vegetable concoctions was never before used for any purpose. The amalgamating pans in the mills surpassed the cauldrons of Macbeth's witches in the variety and villainousness of their contents. The millmen poured into the pans all manner of acids; dumped in potash, borax, saltpeter, alum, and all else that could be found at the drugstores, then went to the hills and started in on the vegetable kingdom. They peeled bark off the cedar trees, boiled it down until they had obtained a strong tea, and then poured it into their pans, where it would have an opportunity of attacking the gold and silver stubbornly remaining in the rocky parts of the ore."

Nor did the mining company Merlins stop with cedar. "The native sagebrush, which everywhere covered the hills, being the bitterest, most unsavory, and nauseating shrub to be found in any part of the world, it was not long before a genius in charge of a mill conceived the idea of making a tea of this and putting it into his pans. Soon the wonders performed by the sagebrush process, as it was called, were being heralded through the land.

"The superintendent of every mill had his secret process. Often when it was supposed that one of the superintendents had made a grand discovery, the workmen of the mill were bribed to make known the secret. To guard as much as possible against this, the superintendent generally had a private room in which he made his vile compounds. 'Process-peddlers,' with little vials of chemicals in their vest pockets, went from mill to mill to show what they could do and would do, provided they received from five thousand to twenty thousand dollars for their secret. The object with many inventors of 'processes' appeared to be to physic the metal out of the rock, or at least to make it so sick that it would be obliged to loose its hold upon its matrix, and come out and be caught by the mercury lying in wait for it in the bottom of the pans."

As it turned out, the Comstock ore was finally and effectively treated by crushing, stamping and agitation in a mixture of water, mercury, salt and bluestone (copper sulphate), which speeded and improved the amalgamation process. But many gold and silver ores in the West were "rebellious," in the millman's word, and were so adulterated with sulphur, copper, zinc, lead and other elements that they would not yield to the processes of the Comstock or to ordinary smelting.

In some ores—notably in the Silver King and Santa Rita mines in Arizona—the particles of silver or gold were coated with sulphur compounds which sealed them off so that they would not amalgamate with mercury. As a ditty of the 1860s put it:

Our German fathers, working mines,
First exorcised the devil;
While we affirm that sulphurets
Are the sole cause of evil.

To exorcise this particular evil, a number of important chemical processes were devised over the years. Millmen found that these ores could often be treated by roasting and then by steeping in tanks of chlorine gas, which drew off the sulfur and converted the metal into chloride form. The latter was soluble in water and could be extracted by a percolation process called leaching. There were other treatments, particularly of silver, that were complex enough to reduce skilled metallurgists to tears and oaths.

It was not until the end of the 19th Century that the harried millmen were finally rescued by a process invented in Scotland. There it was discovered that gold and silver could be dissolved in a very weak (.005) solution of sodium cyanide, and that they could be efficiently recovered by running the solution through boxes or beds of fine zinc shavings; the gold and silver adhered to the zinc, and the two metals could then be separated by heating. Cyanidation, as it was called, made possible the extraction of virtually 100 per cent of the precious metal in the ore. When the process was introduced in the West, millmen everywhere hastened to salvage the precious metal that remained in their great heaps of supposedly used-up tailings. And naturally, George Hearst's Homestake Mining Company was racing ahead of the pack.

In 1899, Hearst's metallurgists built an experimental plant with six cyanide vats to treat the sands in the tailings from one of the mills. The results were so encouraging that the company swiftly built a large plant with fourteen 600-ton cyanide vats, then a second one, until by 1902 it was recovering one million dollars a year from what was supposed to be waste.

At the Comstock, Leadville, Coeur d'Alene, Cripple Creek, Telluride, Tombstone, Chloride, Orogrande and thousands of other mining areas, cyanidation recovered at least one billion dollars that had been thought lost forever. Everywhere small steam shovels could be seen tossing sands into cyanidation plants with their vats of clear, light-green liquid. In the mining camps, the steel cyanide drum from Germany, marked I. G. Farbenindustrie, A.G., became as common a sight as the empty wooden dynamite case. It was a great day for free-lance operators, too. All it took was a shovel, a few old barrels, some cyanide and some shredded zinc. And for a little while at least, moribund mining towns felt a new breath of life, as thousands of men hurried in to work over the abandoned dumps.

This was truly the richest of all the gold rushes, but it was so quiet and undramatic that the promoters and the headline writers paid no attention to it at all.

The blissful Jim Butler, who found gold at Tonopah, Nevada, in 1900 while pursuing his burro, displays his benefactor for the camera.

Nightmare in a Utah mining town

Whether a man mined for gold or silver or something less precious, his life span was likely to be short. Hundreds of men died in the mines, and accidents were so frequent that they rarely made news. But on May 1, 1900, there occurred a coal mine disaster so great that it captured headlines abroad.

That morning, 312 men were in the No. 1 and No. 4 shafts of the Winter Quarters mine, at Scofield, Utah. At 10:25, those outside heard a low thud.

In No. 4, black powder had somehow exploded, igniting ankle-deep coal dust. The blast shot through to No. 1 and filled it with suffocating gas. In minutes, 200 men died in the two shafts.

"My God, it is awful," a witness said. "Whole families are wiped out and the women do nothing but shriek and wring their hands day and night." At the burial services, the minister told the widows, "All that made you love them lives and will live forever."

A pile of debris marks the entrance to the Winter Quarters No. 1 mine shaft *(far right)* the day after an explosion in an adjoining shaft killed two thirds of the 312 men who were in the two work areas. Death came so swiftly that some of the mineworkers were found still clutching their tools.

Rescuers stand amid splintered timbers at the entrance to the mine's No. 4 shaft *(background, left)* where the explosion originated. It took 20 minutes of work to clear enough debris to find the first bodies.

Boots and work clothes of miners killed in the disaster lie outside the Edwards boardinghouse, which was turned into a makeshift undertaker's parlor. Practically all of the victims were prepared for burial there.

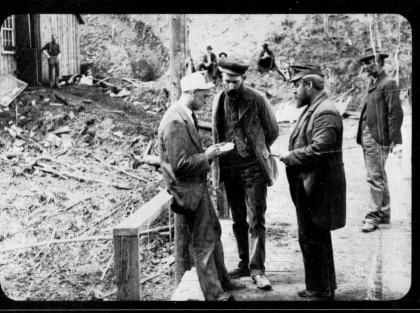

Bodies are loaded into a boxcar for transport to the schoolhouse in Scofield, half a mile away, after the Edwards boarding-house and meetinghouse of the Knights of Pythias could hold no more casualties.

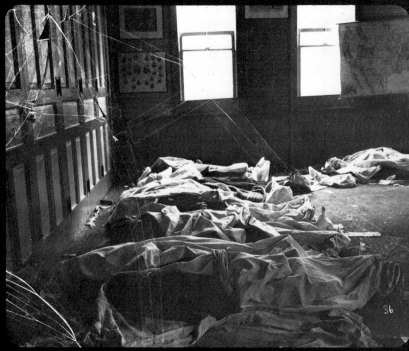

Covered with sheets, some of the 61 men carried to Scofield lie in repose in the schoolroom. One of the witnesses observed that many of their children had celebrated May Day in this room just the day before.

A freight car carrying a special order of coffins is unloaded at the general store in Scofield. The storekeeper ordered all the coffins in Salt Lake City, and still had to send to Denver for enough to bury all the dead.

Some of the flower-bedecked pine boxes rest in the Knights of Pythias hall before burial. School children in Salt Lake City picked and sent three carloads of lilacs, pansies and violets to the victims' families.

A grieving Mrs. John T. Jones and her seven children wait for a wagon to take them to the funeral at the Mormon church house. The disaster made widows of 107 women, and left 268 children fatherless.

The family of Levi Jones surrounds his casket, which was more elaborate than most. For the better part of a week, wrote an observer, "nearly every home in Scofield had a casket on the porch or a casket inside."

Miners James Naylor *(left)* and Evan Williams both got out alive. Williams was close enough to the entrance of No. 1 to race outside before the gas got him. News accounts did not detail Naylor's escape.

Not all the dead miners were from Scofield: here the relatives of those who would
be buried in other Utah towns wait with
the caskets for a specially dispatched funeral train to arrive from Salt Lake City.

Services are held at the Scofield cemetery for 149 miners. After the mass funeral, one man sadly noted that his dead brother-in-law and the other men "who went into the bowels of the earth worked, came home, slept and returned to work, scarcely seeing the glory of the sun and the skies."

The tables start to fill up in one of Leadville's numerous gambling halls around the turn of the century. The game at right is faro.

4 | "The orneryest place this side of hell"

"In the evening," Charles Francis Adams of Massachusetts noted in 1886, "we saw Leadville by gaslight—an awful spectacle of low vice." Leadville might well have regarded Adams' comment as an accolade, for the queen of the mining camps reveled in its reputation as an anything-goes town.

Just a few years before, the local *Chronicle* boasted that the community of 14,000 had 35 brothels, 118 gambling houses and 120 saloons, not counting 19 that sold only beer. The paper was replete with lurid items from the red-light district, and it lavished copy on the brawls, shootings and con artistry that were so much a part of the miners' world.

Its flair for self-advertisement aside, Leadville was not much different from Virginia City, Cripple Creek, Goldfields, Deadwood or any of the other raw, rich mining camps that sprouted across the West. Returning to the East, a stage driver was asked if Deadwood, for example, was as bad as it was cracked up to be. "Worse," he shuddered. "It's the orneryest place this side of hell. There's no law. The gamblin', drinkin' and fightin' goes on all the time, day and night. You wouldn't know when Sunday comes around if you didn't put it down in a book."

Highjinks in the hard-living mine camps

The ways of a man with a maid be strange, yet simple and tame
To the ways of a man with a mine when buying or selling the same.
                                                                    —Bret Harte

Life along the Western mining frontier aroused delicious thoughts of fraud and larceny in many men who, back East, were moderately honest. No doubt the easy-come, easy-go atmosphere of the frontier was a factor in this. So too was the circumstance that silver and gold ore, even bullion, did not seem quite real. They were the stuff of dreams, and did not begin to take on their hard, private value until they were turned into coins with *In God We Trust, Twenty Dollars* on them. Whatever the case, deception was a constant hazard and a source of raffish glee.

It was highly unusual for a prospector to hang onto a claim until every last bit of precious metal—if indeed there really was any—had been taken out. Sooner or later the prospector would try to sell his claim, and before concluding the transaction he would do his level best to enhance its value by whatever means appeared appropriate. If, for example, his claim was worth only a few dollars a ton, he might obtain a true and properly certified assay indicating that ore from his property was worth several hundred dollars a ton. To do this he merely rummaged through his pile of nearly worthless quartz until he found a few bits of good ore, perhaps no bigger than peanuts, and took them for evaluation to an honest assayer, who was obliged to say that the samples indicated a rich yield. Or the prospector might hunt up a dishonest or overly enthusiastic assayer and secure a glowing report.

In *Roughing It,* Mark Twain wrote that "assaying was a good business, and so some men engaged in it,

MRS. LENONT, reliable spiritual and business medium; information of gold localities by "White Bear," her Alaskan guide. Public circles under spirit control Monday, Thursday, Friday, 8 p. m. Second floor Safe Deposit Building, First Avenue.

A teller of mining fortunes advertises in Seattle.

occasionally, who were not strictly scientific and capable. One assayer got such rich results out of all specimens brought to him that he acquired almost a monopoly of the business. But like all men who achieve success, he became an object of envy and suspicion. The other assayers entered into a conspiracy against him, and let some prominent citizens into the secret in order to show that they meant fairly. Then they broke a little fragment off a carpenter's grindstone and got a stranger to take it to the popular scientist and get it assayed. In the course of an hour the result came—whereby it appeared that a ton of that rock would yield $1,284.40 in silver and $366.36 in gold! Due publication of the whole matter was made in the paper, and the popular assayer left town."

Twain, who from 1862 to 1864 worked as a reporter for the *Territorial Enterprise* in Virginia City, occasionally boosted a mine himself. New claims were recorded daily, he wrote, and it was the custom for the claimants to go to the newspaper office and press free shares upon the editors and writers, in return for which those worthies would invariably find it in their hearts to publish some sort of story.

"They did not care a fig what you said about the property," he reported, "so long as you said something. Consequently we generally said a word or two to the effect that the 'indications' were good, or that the ledge was 'six feet wide,' or that the rock 'resembled the Comstock' (and so it did—but as a general thing the resemblance was not startling enough to knock you down). If the rock was moderately promising, we followed the custom of the country, used strong adjectives and frothed at the mouth as if a very marvel in silver discoveries had transpired. If the mine was a 'developed' one, and had no pay ore to show (and of

A 19th Century artist wryly charts the life of the "Honest Miner": a cycle of luck, drink, poverty, salting his poor holding with gold in order to sell it, succumbing to the "tiger"—the faro dealer—and finally burial.

course it hadn't), we praised the tunnel; said it was one of the most infatuating tunnels in the land; driveled and driveled about the tunnel until we ran entirely out of ecstasies—but never said a word about the rock. We would squander half a column of adulation on a shaft, or a new wire rope, or a dressed-pine windlass, or a fascinating force pump, and close with a burst of admiration of the 'gentlemanly and efficient superintendent' of the mine—but never utter a whisper about the rock. And those people were always pleased, always satisfied. Occasionally we patched up and varnished our reputation for discrimination and stern, undeviating accuracy, by giving some old abandoned claim a blast that ought to have made its dry bones rattle—and then somebody would seize it and sell it on the fleeting notoriety thus conferred.

"There was nothing in the shape of a mining claim that was not salable," Twain went on to say. "We received presents of 'feet' every day. If we needed a hundred dollars or so, we sold some; if not we hoarded it away, satisfied that it would ultimately be worth a thousand dollars a foot. I had a trunk about half full of 'stock.' When a claim made a stir in the market and went up to a high figure, I searched through my pile to see if I had any of its stock—and generally found it."

While publicity was handy, a more direct route highly favored by miners was to salt an unproductive claim with a little of the genuine article. On placer claims the prospector merely had to distribute some gold dust in likely locations and invite the purchaser to try a few experimental pans, clinching the sale when the sucker found color.

It was also possible, and not uncommon, for smart operators to do their salting with a shotgun, called by oldtime miners "the gun that won the West." A shotgun could be loaded with a charge other than buckshot; fine gold dust would do nicely, too, as indicated in reports of a transaction that took place near the town of Columbia in Tuolumne County, California, in the 1850s. It happened that a group of American miners were stuck with a large but poorly paying claim in the bottom of a gulch, while nearby a company of Chinese gold panners were doing very well. Naturally the American miners wished to sell their rotten real estate to the Chinese, first salting it, but they were uncertain as to exactly how to proceed. Their claim was much too large to be salted everywhere; yet somehow the gold dust had to be placed where the Chinese, while

A stock certificate issued by the Nevada Gold and Silver Mining Company in 1874 may have been less than gilt-edged by 1876; in at least tv

they were taking sample pans, would be sure to find it.

At length one of the Americans decided that it would be possible to salt the claim while the Chinese were actually looking on, and to that end he killed a large snake and gave it to one of his partners, who was instructed to remain out of sight behind a sandbank overlooking the claim. At a signal the snake was to be slid over the sandbank to land at an indicated spot. Thereafter the scheme would go easily enough, for the American would be walking near the Chinese with his gold-loaded gun under his arm.

According to a witness, four Chinese sampled the earth in various parts of the claim but found no color. "Well, John," said one of the Americans, "where you try next? Why not try that corner, over there?"

The Chinese were instantly suspicious. They were familiar with salt. "No likee dat corn. Tlie him nudder corn," said one, pointing in the opposite direction. Thereupon the American behind the sandbank deftly tossed the dead snake, which glided down and came to rest in the place where the Chinese had pointed.

Kaboom! Boom! The ground was shot full of rich dust and the Chinese were properly grateful to their hosts for having destroyed the reptile. They struck

gold and bought the claim, allowing the Americans to take off for better diggings elsewhere.

The same techniques, with some variations, worked equally well in hard-rock mines. The standard method of hornswoggling a prospective purchaser was to get hold of a few bushels of rich ore from a good mine and scatter it at the bottom of the shaft or along the drift of a poor one. "Why," the happy purchaser would say, "this stuff looks as good as the $500-a-ton rock from the Golden Calf mine," little realizing that it actually was. However, this basic technique was effective only in the early days of the mining boom, before purchasers became smart and wary enough to enlist the services of professionals in appraising the property.

Still, there were always a few eager boobs who were sure they could trust their own judgment. In the 1860s, during a period of intense trading in shares of mines on the Comstock, promoters salted a worthless hole called the North Ophir. They cut silver half dollars into pieces, pounded them into lumps, blackened them and mixed them with barren rock at the bottom of a short, economical shaft. Although silver rarely occurs in nugget form, suckers pounced on the stock of the North Ophir and drove up its price in a fine flurry

ASSESSMENT RECEIPT.

San Francisco, May 18th 1876

Received from Alf Doten the sum

of Ten Dollars,

for Assessment No. 2, Arvada Gold of 10¢ per Share,

levied on the Capital Stock of the PHENIX SILVER MINING COMPANY.

March 30th 1876. The above payment being in full for the proportion of the said Assessment, chargeable to Certificate

No. 1st & 359 for 100 Shares.

$ 300

Jos. Mayrisse Secretary.

Glynas & Dutton, Stationers, 402 and 404 Sansome St.

rters of the year, the assessment receipt from the same enterprise indicates, stockholders were dunned for additional working capital.

Work clothes hang on racks in a changing room in a Nevada mine. Making miners change clothes cut down on their chances to pilfer choice ore — but it also benefited them by reducing ailments caused by their rapid ascent from the torrid depths to the frigid mountain air in sweaty clothes.

that lasted until someone chanced to read "ted States of" on one of the lumps.

Salting by shotgun, sometimes by large-caliber pistol, was also possible in hard-rock mines. If a man in a hard-rock mine stood back far enough, and if the gun barrels had plenty of flare, the whole face of a barren drift could at reasonable cost be blasted with particles of gold. But shotgun salting did require a decent restraint. A man simply could not, without arousing skepticism, blast *any* sort of dust into the face of his drift. It had to be dust that could reasonably have got there in the normal course of geological events.

Thus in British Columbia an engineer came very close to buying a nice little mine that seemed to have at least $80,000 worth of ore in plain sight, with promise of more nearby. Indeed the mine seemed so attractive that the engineer could not understand why the owners were anxious to be rid of it. Accordingly he put some of the ore under a microscope and found that the isolated particles of free gold were deeper in color, redder, than the naturally occurring local product. When he ran a bullion assay on the particles, he found they contained exactly 916.66 parts of gold per 1,000, the rest copper with a trace of silver. This was, not by chance, precisely the formula of the Royal Mint, which used copper to make gold hard enough for use in coinage. The engineer refrained from accusing the owners of having filed gold sovereigns to obtain the particles; he just went on his way without buying.

Prospective buyers could insist, if they suspected chicanery, on having sellers blast out the faces of drifts in quartz mines to expose virgin rock instead of artificially improved stone. But though they watched the holes being drilled and observed the blast, albeit at a safe distance, buyers could still be fooled. Salters merely put gold dust into dynamite sticks, and the explosion drove it into the rock in a convincing way.

The addition of gold dust or filings was by no means the only way of enriching low-grade ore. Solutions of liquid silver nitrate or soluble gold chloride, poured into cracks or drill holes in the rock, raised the assay value handsomely. One group of Easterners, preparing to buy a mine in Dakota Territory, were bamboozled in an even simpler manner. While they watched in amusement, a drunk staggered up to a pile of samples from the mine, tripped and smashed his whiskey bottle on the pile. It was not until much later, after the mine had been purchased and had proved barren, that the Easterners realized that the drunk had been cold sober and that the bottle had contained gold salts.

Gold salts, in the 1870s and 1880s, were a common ingredient in patent medicines sold for the relief of kidney complaints arising from excessive consumption of booze, an occupational hazard among miners. There were smart lads who, according to more or less reliable reports, used to dose themselves with gold salts to soothe their ravaged insides, and then thriftily urinate on their claims to pass along the benefit.

So bedazzling was the very thought of gold that a good many otherwise sensible men were deceived simply by tall tales. In Calistoga Springs, California, a hotel proprietor, anxious to increase business at this resort, announced that he had found the local water so rich in small particles of gold that he could filter several dollars' worth from a single large barrelful. Now, it was well known that gold did sometimes appear in an extremely fine powder, called flour gold; the particles were so tiny that they would actually float on water. Might the hotelkeeper be telling the truth?

The man was of course lying through his teeth, but there were trusting souls who believed him; and it required an expert to demolish the story. In reply to the announcement Mark Twain wrote this letter:

"I have just seen your dispatch from San Francisco in Saturday evening's Post. This will surprise many of your readers but it does not surprise me, for I once owned those springs myself. What does surprise me, however, is the falling off of the richness of the water. In my time the yield was a dollar a dipperful. I am not saying this to injure the property in case a sale is contemplated. I am saying it in the interest of history. It may be that the hotel proprietor's process is an inferior one. Yes, that may be the fault. Mine was to take my uncle (I had an extra one at that time on account of his parents dying and leaving him on my hands) and fill him up and let him stand fifteen minutes, to give the water a chance to settle. Well, then I inserted an exhaust receiver, which had the effect of sucking the gold out of his pores. I have taken more than $11,000 out of that old man in less than a day and a half.

"I should have held on to those springs, but for the badness of the roads and the difficulty of getting the gold to market. I consider that the gold-yielding water is in many respects remarkable, and yet no more remarkable than the gold-bearing air of Catgut Canyon up there toward the head of the auriferous range. This air, or this wind — for it is a kind of trade-wind which blows steadily down through 600 miles of richest quartz croppings — is heavily charged with exquisitely fine, impalpable gold.

"Nothing precipitates and solidifies this gold so readily as contact with human flesh heated by passion. The time that William Abrahams was disappointed in love he used to sit outdoors when the wind was blowing, and come in again and begin to sigh, and I would extract over a dollar and a half out of every sigh.

"I do not suppose a person could buy the water privileges at Calistoga now at any price, but several good locations along the course of Catgut Canyon gold-bearing trade-winds are for sale. They are going to be stocked for the New York market. They will sell, too."

Although Mark Twain was known to commit an occasional exaggeration, he was quite right when he said, "They will sell, too." During the mining rushes Americans seemed willing to buy almost anything that promised a quick, rich return in silver or gold. In particular they bought, on mining exchanges in San Francisco and Virginia City, huge amounts of stock that was either worthless or subject to manipulation by shrewd corporate directors. For all practical purposes there was no regulation of the sale of securities in the 19th Century. Anyone who printed stock certificates and recruited salesmen could, and often did, make a fortune.

At the lowest and most obvious level of stock swindling was the "assessment mine," which took its name from a prevailing custom of the day. A man who bought stock in a mining company, even though he paid the full price of the shares in cash, was still subject to demands for more money by the company's directors. If the latter decided that an extra $50,000 or so was needed for new machinery or exploration, they levied an assessment on each share of stock — and the stockholders, who were responsible for raising new capital, either paid up or lost their shares.

If the directors of a mine were honest and prudent, and the mine still lost money, too bad. That was the hazard of the game. The famous Bullion mine, located in the heart of the Comstock with bonanzas on both sides of it, assessed its stockholders a total of more than three million dollars and never produced a ton of millable ore. Sometimes, however, the directors or promoters used the device of assessment not to raise capital needed for the mine, but merely to bleed the stockholders. Most small investors were absentees who were unlikely to visit the mine—and even if they did, were not knowledgeable enough to know a rattail from a Cornish pump. Thus when a smart operator got hold of a played-out property, he set up an assessment mine and sold shares in it. He could live comfortably by mailing out requests for money, for months or even years, until stockholders' cash and gullibility were exhausted. And even if stockholders managed to catch up with the swindler and hale him to court, laws dealing with fraud were too weak to afford them much recourse.

Stockholders could also be bilked by use of inside information by owners and superintendents. In 1878, historian John S. Hittell wrote: "In a rich mine the quantity and quality of the ore produced must be regulated by the desire of the directors to buy and sell. The rich deposits were concealed when the stock was

to be bought up, or worked with every energy when it was to be thrown on the market. The superintendent of every prominent mine conducted on such principles had his book of ciphers, so that he could send secret messages to his masters and let them know whether the ore was growing richer or poorer, enlarging or diminishing in quantity. Every trick that cunning could devise to make the many pay the expenses, securing to the few the bulk of the profit, was practiced on an extensive scale, in the most active of all the stock markets. On such a basis not less than a dozen of the millionaire fortunes of San Francisco have been built."

While reporting on the Comstock, Dan De Quille described the secrecy of operations in his history of the lode, *The Big Bonanza*. In a large mine the workers were deliberately and carefully segregated—those who worked on one level were forbidden to visit another, and even on their own level were not permitted to wander far from their assigned area. The men had "little opportunity of knowing when a development has been made at a particular point in a mine, or anything about the value of any body of ore that may be encountered.

"When a crosscut is being run at a point where it is thought that ore will be found, the work is carried

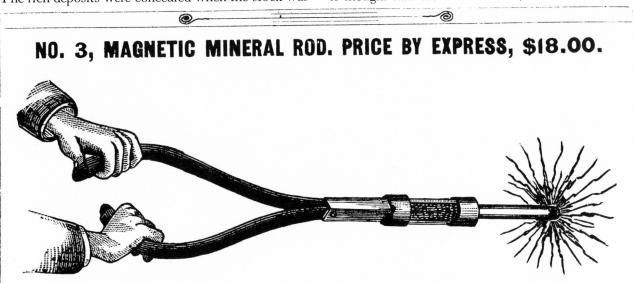

NO. 3, MAGNETIC MINERAL ROD. PRICE BY EXPRESS, $18.00.

THIS instrument is the strongest and most powerful instrument made for the purpose of hunting hidden treasure.
It is known to trace a hundred yards or more easy. Location of veins of gold being so accurately demonstrated that experts were convinced of its merits.

A magnetic dowsing rod in a 1901 catalog may not have found gold, but it had power to call forth flights of golden fancy from a copy writer.

on by what is called a 'secret shift.' This shift is composed of the oldest and most trustworthy men in the mine — men who will work for weeks in a drift that sparkles with native silver and yet remain as mute as oysters when aboveground. These secret-shift men generally find their silence profitable. They are helped to a few shares of the stock at the low figure at which it is probably selling when the ore is found, and pocket whatever advance there may be when the nature and extent of the new development have been made known."

"A few shares" could be extremely profitable. After a bonanza was found in the Crown Point mine in the early 1870s, the stock rose from $3 to $1,825 in 21 months. In the adjacent Belcher mine, shares went from $1.50 to $1,525 between September 1870 and April 1872. A month later the bubble had burst, but for those who had inside information the rewards were great. "The men working on a secret shift are not sworn to secrecy," wrote De Quille, "and it is seldom that they are even pledged — they know why they are selected and what is expected of them. When a secret has been divulged and the guilty person cannot be discovered, every man on the shift is discharged, and not one of them will again be employed on a secret shift in any mine until the real culprit has been found."

Mineowners and superintendents were not always content to rely on bribery or threats to ensure secrecy. According to a report written in 1883 for the United States Geological Survey, on the Comstock "the custom grew up of confining the men at work for days below the surface. The object of this confinement was evident. To imprison the miners was the best guarantee that they would not make premature disclosures; so the trustees became jailors for the time being. The miners so held did not usually grumble, as they were fed and cared for with particular attention, and their wages were often temporarily increased. If all stockholders in a mine had been apprised daily of the developments thus made, the practice might not have been objectionable, but as a matter of fact many were kept ignorant by the willful neglect or refusal of the trustees to accord the information desired."

The practice of imprisonment began on a small scale in 1863, but by 1868 as many as 25 men were confined for three days in one mine, and in 1869 in another, 18 men were held underground for a week. Anger began to rise among workers and their families, and when in 1872 the superintendent of the Ophir mine imprisoned four men for three days in the steaming-hot 1,100-foot level, he was threatened with writs of habeas corpus and lawsuits for unlawful detention. That brought the practice to an end.

Outsiders or even frustrated stockholders trying to obtain information about their own investments haunted the saloons of Virginia City, Leadville and other Western mining towns, offering free drinks and cash in return for helpful hints from off-duty underground workers. Occasionally the information seekers showed considerable ingenuity, as Dan De Quille reported. At one point, when a rich strike was rumored in a major mine, "every avenue to the lower levels was closed to the outside world. The superintendent was exceedingly close-mouthed and mysterious; the miners were reticent and unbribable. And nothing of the slightest value could be bored, pumped or gouged out of anybody or anything, and finally all the news-gatherers but one drew off and gave it up as a bad job. One man still lingered, day after day, all eyes and ears.

"At last a bright idea struck him. The superintendent came to the mine and, as usual, went down into the lower levels. Our man remained loitering about the works until he came out — lingered until he had seen him take off and throw aside his muddy boots, his clay-besmeared overalls and shirt, and till he had finally taken himself off. Watching his chance, the hungry reporter of mining news darted into the dressing-room, and with his jackknife scraped from the boots, overalls, felt hat, shirt and everything all the mud, clay and earth sticking to them. Of this and the loose particles of ore found in the pockets of the shirt he made a large ball, which was composed of a general average of the bottom, top and sides of the drift run into the new deposit; he had a little of everything the superintendent had touched, and this ball he had carefully assayed. By the result obtained he became satisfied that a strike of extraordinary richness had been made. He immediately telegraphed to his employers in San Francisco to buy all of the stock they could get, and the stock soon went up from a few dollars to high in the hundreds."

Manipulators of mining stocks were never inclined to use the word theft in reference to their operations.

The irregular transfer of gold from one man's pocket to another's, unless accomplished at gunpoint, was regarded as more or less legitimate business by all but the immediate victims. It was not easy to draw a line between crime and free enterprise. In mining camps and settlements, for example, where coins were scarce and paper money never used, it was the practice for men to pay for their purchases in raw gold. Merchants, bartenders and even sporting ladies had balance scales on which payments for large orders could be weighed. For small change, a pinch of dust was extracted by the seller from the buyer's poke. In most camps a pinch was recognized as a dollar, and in the long run big pinches and little pinches were expected to average out.

Was it criminal, unsporting, sharp practice, or merely smart for a merchant to let the nails of his thumb and forefinger grow long, and to keep them well manicured? What could be said of bartenders who kept their hair thickly greased, rubbed their fingers through it after each transaction, and shampooed out pay dirt every night? Or of men who, between pinches, constantly squeezed small pebbles to create profitable dents in thumbs and index fingers? And who could blame the poor miners for protecting themselves by adulterating their gold dust with filings of yellow brass? A modest dosage of the stuff could give a miner a 10- or 15-per-cent bonus when he opened his poke for a pinch, or emptied it onto the scales.

Beyond these perplexing philosophical questions lay a far greater one, at least in terms of the money involved. It was called "high-grading." In hard-rock mines gold was seldom distributed evenly throughout a vein. Most areas had only a thin, though perhaps quite profitable, lacing of it. But in certain locations it was naturally concentrated in very rich pockets — in the Frances Mohawk mine in Goldfield, Nevada, at a depth of 200 feet, there was an eight-inch-thick seam that assayed $250,000 a ton. When a miner, laboring in the dark and damp for $3.50 or $4 a day, encountered such a pocket, temptation naturally arose. Ore of that quality was worth $125, or a month's wages, per pound. If the hard-working miner chanced to allow three or four pounds of it to fall into the bottom of his lunch pail, and if he forgot to empty it out and carried it home, he was said to be high-grading. Lamentably, a great many Americans and quite a few Cornishmen were accused of this. Indeed in one California mine there was a Cornishman who was said to be so adept at locating chunks of high grade in the gloom, merely by feel, that he was known as Old Velvet Thumb.

Although the lunch pail was the most convenient container in which to carry rich ore out of a mine, it had drawbacks. The shift boss or the superintendent might accidentally tap the pail with a stick, and if it emitted a deep, clunking sound he might find some pretext to look into it. Miners therefore equipped themselves with leg pockets, which were long canvas tubes suspended inside trouser legs; vest-like garments, containing many pockets, worn beneath the shirt; false- or double-crowned hats, in which as much as five pounds of high grade might be hidden; and various belted and suspended harnesses invented by wives who were handy at sewing. With these aids a strong man could bring home an impressive pay load, although in time he tended to become bandy-legged.

After a miner extracted his high grade from his place of work, it behooved him to sell it immediately, or at least to grind it up. Gold ores — almost all rich, metallic ores, for that matter — were individual in appearance. A practiced geologist or assayer could tell from what mine, sometimes even from what level within that mine, a given piece of ore had come, and thus the high-grader was uneasy until he had obliterated the evidence. As a rule the high-grader sold it to a crooked assayer, who paid about 50 per cent of its value, concentrated it, and fenced it off to a gold buyer who did not inquire into its pedigree.

The mineowners, who regarded high-grading as nothing less than larceny, took steps against it — particularly in Goldfield and in Cripple Creek, the two areas that produced the most highly concentrated ores in the United States. They searched lunch pails and built "change houses," buildings where men going off shift were expected to strip and change their muddy work clothes for clean ones before going home. The owners even went so far as to install jump bars over which naked miners were required to leap, shaking loose any gold that they might be hiding between their legs. But high-grading was never completely eliminated, largely because the miners regarded it as their natural right, a heaven-sent fringe benefit for which they (and their unions) were willing to fight. The low-wage

Soapy Smith and his traveling carnival of crime

Con artist Soapy Smith celebrates July 4, 1898, with some henchmen in his Skagway bar; four days later he was killed in a gunfight.

The king of the mining camp rogues was a transplanted Georgian known as Soapy Smith, who fleeced prospectors clear across the West from Colorado to Alaska. Born Jefferson Randolph Smith Jr., Soapy turned up in Leadville in 1885, and soon earned his nickname by displaying a dazzling mastery of the soap game.

Before a crowd, Smith would flash a roll of bills, ranging from $1 to $100. He would wrap—or appear to wrap—the bills around bars of soap, cover them with paper and toss them in a basket. Anyone could pick a bar for $5; if he was lucky he might win $100. But so dexterous was Soapy that no one except his shills ever came up with more than $2.

Moving on to Denver, Soapy became the acknowledged leader of that boomtown's underworld. Some of his best allies were barbers, who would, while sprucing up a newcomer, assay his wealth; if he appeared loaded, the barber cut a "V" into his hair in back —whereupon one of Soapy's men would lure the easy mark to the Tivoli Saloon and Gambling Hall, there to part him from his money.

By August 1897, Soapy was in Skagway preying on the Klondikers. One of his more imaginative swindles was a phony telegraph office, which charged $5 for messages home; the Klondiker always got a return message—collect, of course.

Soapy's downfall came when some of his henchmen brazenly robbed a miner of almost $3,000 in gold. A delegation of Skagway town fathers demanded restitution. Soapy refused, and in the ensuing showdown was shot to death by the town surveyor.

For all his predatory nature, Soapy was an easy touch for anyone down on his luck, and he donated to the churches. But that was as far as he went in quieting his conscience. "The way of the transgressor," he often liked to say, "is hard—to quit."

scales in the industry, they said, made high-grading almost a necessity.

When an owner caught a miner with his pockets full of "picture rock" and haled him into court, it was well-nigh impossible to get a conviction. The sympathy of the community, which was composed largely of relatives, friends and merchants dependent on miners, was always with the accused man. Judges, who were elected by popular vote, were not likely to side with mineowners — indeed, in Cripple Creek one judge blandly gave the opinion that ore stealing was not really theft because "mineral is real estate" and presumably part of the scenery that belonged to everyone. Unable to get any relief in court, the owners tried to obtain it in other ways. On the night of February 23, 1902, the offices of eight assayers who were suspected of buying high-graded ore were wrecked by explosions of dynamite, widely believed to have been planted by the mineowners. But even that action, although it jolted Cripple Creek out of bed, had little permanent effect.

In Goldfield, where an estimated $30 million was high-graded over the years, many of the mines were operated on a lease basis. The owners of these mines, often absentees who knew nothing of mining and did not care to learn, rented their property for stipulated periods to professionals, who then dug frantically to get out all the gold they could before the leases expired. The system resulted in the heedless gutting of mines that under decent management could have produced more gold in a longer, more tranquil time. It goes without saying that it also provided the perfect climate for high-grading. The leaseholders, acutely aware of the ticking of the clock, dared not object to the miners' exercising their natural right, lest the miners retaliate by staging a slowdown. In one incident at the Mohawk mine a man stepped out of the hoisting cage so overloaded with concealed ore that he fell down. The superintendent, watching, could only say plaintively, "Will someone please help that son of a b----?"

At one juncture, the desperate Goldfield leaseholders and owners resorted to moral suasion. They persuaded a local preacher to come out against the sin of theft. A miner named Frank A. Crampton, who was there, described it thus: "At a Sunday service I attended the preacher extolled to his flock the virtue of honesty and denounced the sin of stealing. I had visions of high-grading becoming something of the past and being ended forever as his flock, deeply attentive, listened in astonished surprise. However there was obvious relaxation, and a sigh of unrestrained relief, when he closed his sermon by adding: 'But gold belongs to him wot finds it first.' I admit I was somewhat relieved myself."

Not all high-grading was done by $3.50-a-day miners. Thomas A. Rickard, a leading mining engineer of the day, wrote a wry little study called *Rich Ore and its Moral Effects* in which he pointed out that high-salaried executives also practiced the art. "I know of the president of a mining company (not at Goldfield but in a neighboring district) who used to visit the mine whose operations he supervised and on the occasion of each visit he filled his dress-suit case or valise with specimen ore. Finally, at the end of the fiscal year he told the manager to debit him with $4,000. The manager told me that he had kept tally on the little shipments that went from the mine in the president's valise, and he estimated the total at $22,000."

Morality on the mining frontier was always refreshingly relaxed. In that exuberant world, all the vices and many of the sins so roundly condemned back home in the blue-nosed East were not only tolerated but enthusiastically practiced. Gambling was more prevalent in 19th Century silver and gold camps of the West than at any other time or place in American history. Men who had struck it rich yesterday, or had hopes of doing so tomorrow, had an unbounded optimism that seemed to compel them to stake their fortunes, large or small, on unlikely and usually losing propositions.

Professional gamblers, honest and dishonest, hurried to set up their tables or planks close to the scene of every new strike, and speedily relieved the miners of their heavy burden of dust. Albert D. Richardson, who went to Denver in 1859 as a reporter for the Boston *Journal*, found the town a hodgepodge of cabins clustered around the Denver House, "a long low one-story edifice, 130 feet by 36 feet, with log walls and windows and roof of white sheeting. In this spacious saloon, the whole width of the building, the earth was well sprinkled to keep dust down. The room was always crowded with swarthy men armed and in rough

A patron in a Virginia City saloon pours his own, as was the custom of the day. He was certain to fill his glass to the brim, for the price was the same however much he poured. Drinking went on from morning until late at night; it was not unusual for a man to down a quart of whiskey a day.

costumes. The bar sold enormous quantities of cigars and liquors. At half a dozen tables the gamblers were always busy, day and evening. One in woolen shirt and jockey cap drove a thriving business at three-card monte, which netted him about one hundred dollars (almost a pound of gold) per day. Standing behind his little table he would select three cards from his pack, show their faces to the crowd, and thus begin:

" 'Here you are, gentlemen; this ace of hearts is the winning card. Watch it closely. Follow it with your eye as I shuffle. Here it is, and now here, now here and now' (laying the three on the table with faces down) 'where? If you point it out the first time you win; but if you miss you lose. Here it is, you see' (turning it up); 'now watch it again' (shuffling). 'This ace of hearts, gentlemen, is the winning card. I take no bets from paupers, cripples or orphan children. The ace of hearts. It is my regular trade, gentlemen—to move my hands quicker than your eyes. I always have two chances to your one. The ace of hearts. If your sight is quick enough, you beat me and I pay; if not, I beat you and take your money. The ace of hearts; who will go me twenty?' "

Sometimes gamblers and saloonkeepers were so anxious to get their hands on the gold that they all but climbed down into the holes beside the miners. In Oro City, Colorado, the richest pocket in the placer diggings was discovered and staked in 1860 by two veteran miners named Jack Ferguson and Pete Wells. On good days they each took several ounces of gold from their claim, but because Ferguson was a robust drinker and Wells a dedicated bettor, they kept none of the money. A smart grogseller and gambler opened a saloon immediately beside their location, in order to get first crack at the gold, and efficiently stripped them of it. From the gambler's point of view Ferguson and Wells were merely amiable, trained groundhogs who labored for him all day while he took his ease. When the claim played out he went on his wealthy way while the miners drifted off dead broke.

In any case, miners were so fond of gambling that they did not need the services of a professional to help them go broke. A man named Chauncey Canfield described an exemplary game at Coyoteville, California: "There were four partners in one of the richest claims on the hill and they got to gambling together. They

started in playing five dollar ante. Then they raised it to twenty-five dollars ante, and Jack Breedlove, one of the partners, cleaned out the rest of them, winning twenty-two thousand dollars. Not satisfied with this they staked their interests in the claim, valuing a fourth at ten thousand dollars, and, when the game quit, Zeke Roubier, another partner, won back eight thousand dollars and held on to his fourth interest. The other two went broke and Breedlove ended by owning three fourths of the claim and winning fourteen thousand dollars, so that he was thirty-four thousand dollars ahead. He offered his old partners work in the mine at an ounce a day, which they refused, packed their blankets and started out in search of new diggings."

Women were so scarce on the mining frontier that any female was treated with a respect that sometimes verged on idolatry. Caroline Leighton, a highly proper New England lady who visited the Pacific Northwest later wrote: "Among the miners of the upper country, who had not seen a white woman for a year, I received such honors that I am afraid I should have had a very mistaken impression of my importance if I had lived long among them. At every stopping-place they made little fires in their frying-pans, and set them around me, to keep off the mosquitoes while I took my meal. As the columns of smoke rose about me, I felt like a heathen goddess, to whom incense was being offered."

Ladies who lacked Mrs. Leighton's refinement were still accorded considerable honor, at least until proper wives and mothers began to turn up in large numbers and drive the rougher ladies into specified red-light

A RIP-ROARING GOLD CAMP FOLLOWER

Calamity Jane Cannary, known to many Western miners as an affectionate but eccentric harlot, claimed to excel at marksmanship and other rough pursuits. Calamity strode about the gold camps toting a rifle and a brace of revolvers, habitually wearing men's clothes, not the buckskin costume she affected in the carefully posed picture at left. Nor did she always maintain her figure in the best shape. One observer said she was "built like a busted bail of hay." She boasted of unproven triumphs as a bull whacker, wagon-train boss, Indian scout and Pony Express rider. But what Calamity Jane really did mostly was brawl, curse, booze and roister with the best of them—including Wild Bill Hickok, who was a barroom friend in life, and next to whom she was buried after she died, a pauper, near Deadwood in 1903.

districts and, eventually, clean out of town. But until they were overborne by the forces of uplift, prostitutes and madams, with their business-like insistence on reasonably civilized behavior, brought elements of gentility to mining towns. For many men a parlor house was not only a place of entertainment but a club and confessional. Madams and prostitutes often grubstaked miners, or contributed to the general welfare by investing in legitimate businesses.

A number of the prostitutes were piquantly named— the Irish Queen and the Spanish Queen, Little Gold Dollar, Molly b'Damn, Em' Straight-Edge, Peg-Leg Annie and Contrary Mary. (The names of the customers of these ladies also were not without distinction: Jack the Dude, Johnny Behind the Rock, Coal-Oil Georgie, Jimmy the Harp and Senator Few Clothes.) Moreover the reputations of the ladies were adorned with sentimental tales that helped to promote the legend of the Whore with the Golden Heart. According to one miner who knew her, the Irish Queen was ruthless in extracting money from her clients but when they fell on hard times she would wade through snow in midwinter "to take soup to some poor devil to whom she didn't owe a damned thing. Just a heart of gold and nothing else!"

Molly b'Damn was described by an Idaho contemporary as "an uncommonly ravishing personality. Her face gave no evidence of dissipation, her clothes no hint of her profession. About her, at times, was an atmosphere of refinement and culture." Occasionally "she quoted with apparent understanding from Shakespeare, from Milton, from Dante." Another prostitute named Mollie May, who worked in Leadville in the 1870s, so impressed the miners with her kindness and gentility that when she died in 1887 the *Evening Chronicle* printed a poem that went in part:

Talk if you will of her
But speak no ill of her—
The sins of the living are not of the dead.
Remember her charity,
Forget all disparity;
Let her judges be they whom she sheltered and fed.

In Virginia City, the reigning prostitute was the legendary Julia Bulette. Reputedly of Creole origin, she turned up when the town was still only a collection of

153

Dawson City's ladies of pleasure await visitors in their muddy, tightly packed red-light district. Characterized in 1898 by a harsh observer as "strident harpies with morals looser than ashes," the women had been forced from the town proper into this stark row of cribs across the Klondike.

Julia Bulette, Virginia City's favorite lady of the evening, shows off a gift from her admirers in the fire company. Not long after, the man in the inset, a ne'er-do-well named John Millian, strangled her for her jewels.

tents, saloons, flimsy rooming houses and cabins, and soon she set up a parlor house that became a center of elegance. Julia was so bewitching that she could command prices as high as $1,000 a night for her company. She served fine wines and a delightful French cuisine, and almost daily adorned her house with fresh flowers rushed from the Coast by the Wells, Fargo and Company express.

To the lonely miners Julia's house was indeed a home, filled with touches of long-forgotten grace, and they were extremely fond of her. They made her an honorary member of the Virginia City Fire Company and on the Fourth of July, 1861, so goes the story, she led a parade through the town, riding in a fire truck and carrying a fireman's trumpet filled with roses, while her proud "boys" marched in red-shirted ranks behind her. For ordinary, day-to-day transportation she used a handsome carriage which had on its doors the Bulette crest: four aces surmounted by a lion couchant.

When hundreds of miners became ill from drinking polluted water Julia turned her house into a hospital and herself into a nurse. During the Civil War she raised large sums for the Sanitary Commission, the Red Cross of its day. But such good works were not sufficient to preserve her social position when proper ladies and gentlemen moved into the community. Her boys continued to love her, but she could no longer sit in the orchestra of the theater surrounded by a swarm of admirers. She was forced to retreat to a box at the side, partially curtained to protect her from the glares of the saintly.

In 1867, on a cold winter morning, Julia was found strangled in her bed, apparently by a robber who had made off with her jewelry. Her funeral was one of the most impressive Nevada had ever seen. All the mines on the lode shut down while hundreds of men, led by the firemen and the Metropolitan Brass Band, followed her body to its grave in unconsecrated ground and wept unashamedly when the musicians cut loose on the way home with the gay, marching strain of *The Girl I Left Behind Me.*

Four months after Julia's death a French adventurer named John Millian, recently jailed on an unconnected charge of attempted murder and robbery, was found to have some of her jewelry in his possession. He was immediately accused of her murder and tried and convicted. So many people wished to attend his hanging that it had to be staged in a large natural amphitheater in the hills north of town. Once again all the mines on the Comstock shut down while Julia provided Virginia City with another great holiday.

In addition to their enormous influence, the successful madams of the mining frontier were all good businesswomen. The great Mattie Silks of Denver, who came there from Kansas in 1873 and operated parlor houses for many years, in her old age gave an interview to a reporter who wrote: "She defended calmly but without emotion the life she had led. And she said at the first, 'I went into the sporting life for business reasons and for no other. It was a way in those days for a woman to make money and I made it. I considered myself then and do now—as a business woman. I operated the best houses in town and I had as my clients the most important men in the West.

"'I never took a girl into my house who had had no previous experience of life and men. That was a rule of mine. Most of the girls had been married and had left their husbands—or else they had become involved with a man. No innocent young girl was ever hired by me. Some of my girls married the customers. And my girls made good wives. They understood men and how to treat them and they were faithful to their husbands.'"

But while the madams prospered, the ordinary prostitutes did not—or at least did not keep the gold that they earned by the bucketful in their short careers. Most of them were ignorant, raucous women who died early. Crude abortions, alcoholism and other diseases took an appalling toll. Suicide, frequently by gunshot or laudanum, was commonplace. Their obituaries were often short and cruel, like this example from a Virginia City paper of the 1870s: "A woman known as Grace Fanshaw, age about twenty-five, residing at 26 South D Street (in the red-light district), committed suicide last night by drinking laudanum during a fit of despondency brought on by blighted love, acute alcoholism and bad investments."

There was a sad disparity between the myth of the Whore with the Golden Heart and the facts of her life. But as the years passed and miners retreated to their rocking chairs to cultivate their memories, prostitutes were awarded a place of honor not far below mother, the flag and the 10-pound nugget of solid gold.

An elaborate fountain is one of the many ostentations garnishing the lawn of Linden Towers, James Flood's palace in California. It required a 121-man grounds crew to plant and care for the 35-acre estate, studded with statuary, hundreds of gas lamps, banks of violets and walls of roses.

5 | The halls of the mining kings

In their powerful desire for material expression, most mine-made millionaires immediately set about flaunting their wealth with the gaudiest of purchases, the largest invariably being an immense, costly, often extremely ugly mansion.

For Horace and Baby Doe Tabor (*page 179*) the choice in Denver was a villa with a three-acre front lawn that was home to 100 peacocks. For Tom and Carrie Walsh (*page 163*) it was a 60-room palace in Washington, D.C., with four-story-high marble columns and a multicolored glass dome.

But it was Silver King James C. Flood who led the field with Linden Towers, the million-dollar mansion shown here. The furnishings for his 45-room house in Menlo Park, California, included Italian murals, silver faucets and pearl-inlaid billiard cues. The banquet hall boasted a table that could seat 40, and there was a smaller dining room for the children. Flood's huge bedroom suite was furnished entirely in rosewood, and there were 11 only slightly less opulent guest suites.

Linden Towers was five years a-building. When finished in 1879, critics took one look at its tiers of white turrets and towers and scathingly christened it "Flood's Wedding Cake."

Flood apparently did not like it much better, and made it his main home for barely five years before moving to grand new digs on San Francisco's Nob Hill.

JAMES C. FLOOD

WILLIAM S. O'BRIEN

JAMES G. FAIR

JOHN W. MACKAY

Some plain folks who struck it very rich

If a composite sketch were made of men who struck it rich in silver or gold, the picture might be that of an immigrant Irishman or the son of one. Irishmen were no luckier than anyone else, but in the second half of the 19th Century, when the great precious-metal strikes were made, there happened to be a lot of young immigrant Irishmen in the country. Many of them were poor, foot-loose and willing to go West to take their chances in the hills. Other likely characteristics of the composite Croesus were a desire to build an impressive mansion, an itch to go back to Europe to stun the natives with displays of wealth, and a hankering to be a United States senator. In addition, the sketch might show an unlettered but intelligent man, courageous, with a wife as gaudy and assertive as a macaw.

And luck of course was always in the picture, as in the case of a man named Warren Woods from the East, who went to Victor, Colorado, a suburb of Cripple Creek, in the 1890s with his sons Frank and Harry. All they wanted to do was build a modest hotel in the mining town. But in digging the basement they were hindered by quantities of exceedingly heavy material that paid them $30,000 a month when they recognized it for what it was: high-grade gold ore. In time they ran up their holdings to $15 million.

Or consider James Doyle, James Burns and John Harnan, who also made their fortunes in Cripple Creek. Doyle, wandering over the local mining ground, chanced upon a small, unclaimed parcel of land, triangular in shape and covering only one sixth of an acre, that lay between two large mines. It happened some-times that such scraps of free real estate remained unnoticed. Doyle claimed it, and in partnership with Burns and Harnan dug into a vein that assayed $640 a ton in gold. However, they dared not let word of their find leak out. They realized the owners of the adjacent mines would file suits under the federal "apex law," insisting that the vein belonged to them because the apex — or uppermost point — of the vein was located on *their* property, not that of the three Irishmen.

Doyle, Burns and Harnan had no money to hire lawyers to defend their little claim, and without expensive counsel they would have been overwhelmed in court. Knowing that their only hope lay in getting out enough ore, secretly, to build up funds for the legal battle that was surely coming, they put up a shack on their property and dug in the floor at night by candlelight. They kept this up for some time, smuggling out the ore with a mule team and wagon, until the wagon wore a noticeable rut in the earth and they were discovered.

Their neighbors immediately attacked them with barrages of subpoenas and injunctions. But by then the Irishmen had taken $90,000 from their one sixth of an acre and were ready to fight. Indeed, they fought so well, defending no fewer than 27 lawsuits, that they were able not merely to keep their claim but to counterattack weak points in their neighbors' claims. Eventually they accumulated 183 acres that ultimately yielded $65 million.

Although no one ever struck it rich without luck, successful miners also had a cultivated asset — a prepared mind — that enabled them to comprehend good fortune when they saw it. A sound man took the trouble to study basic geology, to learn where gold and silver were likely to be found and what the many forms of ore looked like.

One of the best in this regard was Thomas Francis Walsh of County Tipperary, Ireland, who in 1880

These dour-faced men lived the American success story, rising from humble origins to become owners of Nevada's Big Bonanza mine. A year after the 1873 strike, each of the Silver Kings was a multimillionaire.

was prospecting near Leadville. One day as he glanced at an abandoned cabin in the hills his eye was caught by a glitter on the low, mounded roof of sod and earth. Walsh approached the cabin and saw that the glittering object was a piece of mineralized quartz. Exploring, he found a mine shaft not far away that had been dug by the cabin's vanished builder. But the shaft ended in barren granite. Where had the quartz come from? When he saw no likely source in the immediate neighborhood he entered the cabin, sank his pick into the dirt floor and hit more quartz that looked as though it contained silver. It turned out that the cabin had been built squarely on the apex of a rich little vein from which he was able to take $75,000 in two months.

More than a decade later, while looking at abandoned silver workings at Ouray, Colorado, Walsh noticed some rock that appeared to have gold in it. He hired a crew to sink an exploratory tunnel and soon the men struck an 18-inch streak of very colorful ore that contained lead, zinc and copper pyrites. Below it was a three-foot vein of dull, grayish quartz — and it was the latter that attracted him.

As Walsh wrote, the lower vein "had none of the shining mineral in it and looked so barren that the average miner would consider it no good, but as I examined it closely I saw little specks and thread-like circles of glittering black material all through it, which my experience told me was gold in telluride form. I became so alert in sampling the grayish-looking quartz that my man Andy grew uneasy and asked me not to work so hard. Thinking that I didn't see the low-grade but spectacularly showy metalliferous streak he called my attention to it, saying that that was the pay streak. I was laughing inwardly and said, 'Never mind, Andy, I always assay everything in a vein.' However I took some samples of the lead and zinc streak merely to allay Andy's spirits and got returns of $8 a ton, while the samples from the common-looking neglected quartz ran as high as $3,000." From this played-out silver mine, which he speedily bought, Walsh extracted a fortune of over seven million dollars in gold that enabled his daughter, Evalyn Walsh McLean, to buy several interesting trinkets, including the Hope Diamond.

Women frequently got the profits from rich mines but seldom operated or owned them. A notable exception was Eilley Orrum, the self-styled "Queen of Washoe," who came into possession of a small piece of the great Comstock Lode soon after its discovery. Eilley, a poor girl born in Scotland, was converted at 15 to the Mormon faith by a missionary and in 1842 made her way to Nauvoo, Illinois, to join the assemblage of Latter-day Saints. At 16 she married the Elder Edward Hunter, who was nearly 50, and journeyed with him to Salt Lake. But soon Elder Hunter began to marry other young ladies. Eilley was unhappy and divorced him. She then married another Mormon and accompanied him to the Carson valley not far from the still-undiscovered Comstock Lode. But her second husband had some failings too, and she obtained a second divorce, remaining in the Comstock area while he returned to Salt Lake. She began to cook and wash clothes for the miners in nearby Gold Canyon, and at about the time the Comstock Lode was discovered she accepted from one of them a 10-foot claim in settlement of an unpaid board bill.

The 10-foot claim turned out to be extremely valuable. It was located next to a similar claim owned by Sandy Bowers, another boarder, and soon logic and perhaps love led to a consolidation. Eilley married Bowers and their joint 20 feet began to yield as much as $50,000 a month. Eilley, a superstitious girl who believed she could read the future in a crystal ball, which she called a "peepstone," saw herself as a queen and set out to live as one. She built a $300,000 mansion in Washoe Meadows, some 10 miles from the Comstock, and adorned it with $3,000 mirrors from Venice and lace curtains that cost $1,200 each.

As a queen, Eilley felt the scarcity of other royalty in the neighborhood and decided to call on Queen Victoria. But when she and her husband arrived in London in the late spring of 1862 they found the frosty United States Ambassador, Charles Francis Adams, unwilling to arrange an introduction to the British monarch. Victoria was not in the habit of receiving guests who had been divorced once, let alone twice, and she would make no exception in this case.

After spending $250,000 in the shops of London and Paris dressmakers and jewelers, Eilley returned to Washoe Meadows, carrying with her a cutting of English ivy clipped from the wall of Westminster Abbey. She planted the ivy beside her mansion and proudly pointed it out to the miners as a personal gift from her

Thomas Walsh, a onetime innkeeper, had made millions in gold when he and his wife were photographed in 1904. Among their social conquests were President Theodore Roosevelt and King Leopold of Belgium.

friend, Victoria. That did not seem so unlikely to them —after all, Victoria did not own even one foot of the Comstock, and perhaps all she could afford to give a visitor was a little green plant.

In the late 1860s, after several years of diminishing returns, the 20-foot claim played out, and in 1868 Sandy Bowers died. Eilley left Washoe and for many years worked with her peepstone as a fortuneteller in Reno and San Francisco, where she eventually died in a poorhouse. But before she departed from Washoe she killed the ivy by pouring on it a strong solution of lye. She did not want Victoria's gift to fall into the hands of someone who might not appreciate it.

If Eilley reigned alone as Queen of Washoe, the Comstock had no fewer than four kings—John Mackay, James Fair, William O'Brien and James Flood. The first three came from Ireland to the United States

as children; Flood, the son of newly arrived Irish parents, was born in the New York slums. By 1876, having laid hold of the Big Bonanza, they were famous worldwide as the Lords of the Comstock. Superficially, the four had much in common—all were of scant education and penniless background, and all had gone to California in the early years of the gold rush. But at a closer look the four Irishmen were as diverse as men can be. All that united them was an interest in silver.

Flood and O'Brien were not miners but San Francisco saloonkeepers. As they poured drinks they listened carefully to the talk of their customers, many of whom were traders on the San Francisco Stock and Exchange Board, and they picked up valuable bits of information. In the mid-1860s the exchange was boiling with activity. The stocks of thousands of mines fluctuated at a prodigious rate, sometimes rising or falling as much as 1,000 per cent in a matter of weeks. The two saloonkeepers, particularly Flood, became such adept speculators that they gave up dispensing booze and formed a partnership to deal in stocks.

Mackay and Fair began as placer miners on the Mother Lode. By the time the Comstock was discovered in 1859, they had already accumulated a good deal of knowledge and experience. Fair, indeed, had previously built and operated a quartz mill in the Sierra. And Mackay—though broke and obliged to start his Washoe career as a pick-and-shovel man at three dollars a day — was so able that within a few years he became a superintendent and later a major stockholder in a small but profitable Comstock mine.

Mackay was one of the most admirable men the Western mining industry ever produced—intelligent, honest, courageous, modest and generous. Fair, who was commonly called Slippery Jim, was Mackay's opposite. Some notion of Fair's character can be got from the things the newspapers said of him. A San Francisco newspaper editor, Arthur McEwen, once called him "gross, greedy, grasping, mean and malignant," and his obituaries were among the most savage in the American past. "Since James G. Fair died last week," wrote that same editor late in 1894, "I have yet to hear a good word spoken of him. Never did I meet a man of good intelligence who had dealings with him of any sort who did not detest him." Yet Fair was a miner of rare skill, and a master mechanic as well.

Although they were so different, each appreciated the other's ability, and in 1868 they formed a partnership, hoping to get control of one or more of the larger mines on the Comstock Lode. Flood and O'Brien, the other two Irishmen, were brought into the partnership mainly because of Flood's good business judgment and his dexterity in buying mining stock quietly, without causing a large run-up in price. O'Brien, the least talented of the four, was included because of Flood's loyalty to him. "I would never have anything that Billy didn't share," Flood said, while O'Brien himself later remarked that "I just got hold of the tail of a kite and hung onto it."

By early 1869, after its stock had fluctuated wildly between $7,100 and $41 in the space of 12 months, the four partners had managed to buy enough shares at the lower prices to win control of a mine called the Hale & Norcross, which occupied 400 feet near the center of the Comstock Lode. Although the Hale & Norcross mine had earlier yielded more than two million dollars, it had produced little in the preceding year. However, by good luck and good management, the partners succeeded in getting the mine into production again, and during the next two years were able to pay $728,000 in dividends. With these profits they then took the gamble that led to the Big Bonanza.

A few hundred yards north of the Hale & Norcross were two unproductive, adjacent claims, the Consolidated Virginia and the California, on which more than one million dollars worth of exploratory work had been done with no reward. The claims were described as "literally honeycombed with mine workings to a depth of 500 feet." It was the hunch of Mackay and Fair that no one had yet dug deep enough. For a total investment of about $100,000 the partners bought a 75 per cent interest in the two mines and began to explore the depths. At a point 1,167 feet below the surface they hit the top of the Big Bonanza, which extended downward, widening, until at 1,650 feet it ended abruptly as on a floor. In the words of a report later made for the United States Geological Survey, "No discovery which matches it has been made on this earth from the day when the first miner struck a ledge with his crude pick until the present."

From 1873 to 1877 the two Big Bonanza mines yielded nearly $1.5 million a month, and before they were exhausted in 1897, some $135.8 million was taken from them. The net profit to the Silver Kings varied according to the amount of capital each had originally put up, but at a low estimate it was $10 million to O'Brien, $12 million to Flood, $15 million to Fair and $25 million to Mackay. And since they invested their money shrewdly, these fortunes soon increased, placing the four Irishmen among the richest men in the world of the 1870s, thanks largely to James Flood, the most financially astute of the four partners and, it was widely believed, the one most inclined to engage in shady business practices. San Francisco newspapers often accused the partnership of "working the Comstock from both ends," as the *Chronicle* put it, and Flood bore the brunt of the criticism.

To drive up the price of stock in the Consolidated Virginia and California mines, the firm, under Flood's

Sandy Bowers

Eilley Orrum Bowers

Egged on by her profligate husband, Sandy, who boasted, "I've got money to throw at the birds," Eilley Bowers set out in 1861 to build the grandest mansion in all the West — a Nevada palace where she would reign as "Queen of the Washoe." But when the picture above was taken in 1874, Sandy was dead, their claim had run out and Eilley had taken to giving tours in an effort to save her mortgaged home.

leadership, was said by the *Chronicle* to have "published false and exaggerated reports of the extent and quality of the development. Having effected this, they sold enormous quantities." Then they issued pessimistic reports and when the stock fell, the partners quietly bought back their stock. The device, known to San Franciscans as "Flood's milking machine," was said to have made the firm $50 million in only three years. At a time when there were no stringent federal or state laws against such manipulations, the partnership prospered and no member of it was ever tried, much less convicted, for any offense.

How the four partners handled their millions was as different as the men themselves. O'Brien, the simplest and most genial of the lot, did not flaunt his wealth. After he got rich he spent much of his time in the back room of McGovern's Saloon in San Francisco, drinking and playing cards with a handful of cronies. At his elbow he kept a tall stack of silver dollars from which both friends and strangers were free to help themselves. According to the *Alta California*, O'Brien "had more friends in all walks of life, and fewer enemies, than falls to the lot of most rich men."

His most prized possession was not a diamond stickpin or a pair of fine horses but a silver loudspeaking trumpet given to him when he concluded a term as foreman of one of San Francisco's volunteer fire companies. A lifelong bachelor, he lavished his affection on his two married sisters and his numerous nieces and nephews, and was delighted to give them the jewels, houses and clothes that meant so little to him.

In 1878 at 52, O'Brien, the oldest of the Silver Kings, quietly let go of the tail of the kite and became the first of the Lords of the Comstock to die. He left the bulk of his estate to his sisters, who received about four million dollars apiece. Each of his three nephews —the youngest 11 and the oldest 27—got a comfortable legacy of $300,000. His four nieces—one was nine, and the others in their teens—received like sums, with the explicit direction that the money was theirs alone and could not be touched by future husbands. O'Brien had heard of fortune hunters.

James Flood, the second oldest of the four partners, was much given to ostentatious displays of wealth in his personal life and delighted in building great houses. With his riches he constructed Linden Towers, an im-

Silver Kings John Mackay *(left)* and James Fair *(right)* escort Ulysses S. Grant *(standing fourth from right)* and his party into the Big Bonanza mine in 1879. The former President emerged from the 130° heat in the tunnels exclaiming, "That's as close to hell as I ever want to get."

mense, 45-room palace in Menlo Park *(pages 158-159)*. It was a mass of baroque turrets and gables and surmounted by a 150-foot tower. Its interiors were finished in carved walnut, satinwood, mahogany and embossed velvet. There was a stable for 20 horses and a carriage room as large as a hotel ballroom filled with magnificent coupés, traps and broughams. When Flood drove to his office in the city — his coachmen dressed in plum-colored livery, his horses resplendent in silver harness fittings — people watched his progress with awe. "A hush settles over Montgomery Street," reported the *Chronicle*. "It is thus that Queen Victoria arrives to open Parliament."

Not content with Linden Towers, Flood in 1882 bought a square block on Nob Hill and built a more stately mansion. Although its walls were of sandstone fetched from Connecticut and its parlors were Moorish, East Indian and Louis XV in style, the feature that most intrigued San Franciscans was the $30,000 bronze fence that surrounded the place. A servant who was assigned to polish the fence was never able to complete the job, and wore out his life endlessly shining his way around the block. After Flood's death at 62 in 1889 his mansion became the home of the Pacific Union, the most luxurious of the city's clubs.

Flood's great houses were shared by his wife and his two children, one of them a daughter named Jennie. In 1879 he entertained former President Ulysses S. Grant at Linden Towers and a romance developed between Jennie and Grant's son, Ulysses, Jr. Flood was reluctant to announce an engagement because young Grant was penniless, and would be accused of marrying for money. So he gave the President's son a stake, and told him how to increase it in the San Francisco stock market. The investments recommended by Flood proved to be so good that young Grant reportedly made a profit of $100,000 in six months — but then he took the money and ran. Jennie remained a spinster until her death at 67.

James Fair, the man of the unfortunate obituaries, was exceedingly miserly with his money. He hoarded it, invested and reinvested it and eventually owned 60 acres of the choicest real estate in San Francisco, from which he collected nearly $9,000 a day in rents. Yet when his personal fortune had mounted far into the millions he hastily vetoed the plan of his partners to increase the salary of a valuable employee in the mines. "No, no, no," he said. "A hungry hound hunts best."

In the late 1870s on the advice of a doctor, Fair took a round-the-world cruise to rest his mind and body. Not wishing to travel alone, he invited a friend to accompany him. The friend was pleased to accept, assuming that the millionaire would pay for the trip. Fair allowed him to keep that impression until the last day of the journey, when he handed the man an itemized bill for $6,000, half of their joint expenses.

In 1879, after the Big Bonanza had begun to peter out, Fair suddenly took it into his head to run for the United States Senate. He campaigned against another Comstock multimillionaire, William Sharon, and spent an estimated $350,000 to get himself elected. But he soon grew bored with legislative duties. He made only one speech — in favor of the Chinese Exclusion Act of 1882 — and thereafter lapsed into an almost total silence for the remainder of his six-year term.

However, he did not remain out of the public eye. In 1883 his wife of 21 years sued for divorce on grounds of "habitual adultery." Such a charge against a United States Senator, in those staid and proper times, scandalized the nation. It became front-page news in papers everywhere and scores of sermons were preached on the subject, exhorting the Senate to cleanse its ranks of sin. Fair offered no defense, and a judge awarded his wife what was said to be the largest divorce settlement in United States history to that time — five million dollars in real estate, cash and securities. Fair did not mind losing his wife but parting with five million dollars caused him, as he said, "anguish and despair." During the trial both Mackay and Flood (O'Brien had already died) took Mrs. Fair's side against their partner, and thenceforth he had little more regard for them than for "a brace of rattlesnakes."

In his last years Fair recouped his loss and ran up his estate to a reputed $45 million, one of the greatest fortunes ever made in the West. He died at 63 in 1894, not long after having lent money to a needy friend, foreclosed, and taken away the friend's business and home. When someone suggested to him that this seemed a trifle unkind, Fair said only, "A man who can't afford to lose shouldn't sit in a poker game."

After Fair's death there occurred a legal poker game that enlivened the newspapers for several years. In his will he made a few small bequests to some individuals and charities—less than 1 per cent of his huge estate—and left the balance to his son and two daughters. But presently there appeared a swarm of litigants who claimed to have been his secret legal or common-law wives, fiancées or illegitimate children. None of them was able to extract more than token payment from the estate, but there were so many lawsuits, and so many lawyers were engaged on all sides, that a San Francisco magazine happily summed up the situation thus:

"When the Fair will was published several years ago we remarked that a man of his vast wealth should have been more liberal in his contribution to worthy charities. We gladly withdraw the charge. For it has now become clear that a substantial part of the Fair fortune will be devoted to philanthropy of a particularly heart-warming kind. It will enable nearly a score of rich but deserving attorneys to spend the balance of their lives on the lap of luxury."

The most capable of the Silver Kings, John Mackay, was a soft-spoken man of natural dignity and intelligence, who read the erudite *North American Review* and was intensely fond of music and theater. His fortune was greater than that of his partners but he took no particular joy in it. "The fellow who has $200,000 and tries to make more is only borrowing trouble," he said. Among the troubles that came to Mackay was a loss of interest in his favorite game, poker. At a time when his income was approaching $800,000 a month he could no longer find any excitement in gambling for stacks of $20 gold pieces; he once threw down his cards during a game in Virginia City and wandered off, sadly saying, "I don't care whether I win or lose. When you can't enjoy winning at poker, there's no fun left in anything."

Mackay was alone among the partners in possessing a highly developed sense of philanthropy. During his lifetime he gave away at least five million dollars to various charities. And he dispensed staggering sums in casual handouts. When he walked from his hotel to his office in San Francisco so many people buttonholed him on the sidewalk that he estimated the cost of his walk at $50 a block, which he was always careful to put in his pocket beforehand, in gold. He quietly paid pensions to a large number of old Comstock pioneers, and during slack times on the lode, when many miners

were out of work, Mackay picked up grocery bills that came to $3,000 a month. Over the years, he wrote off as gifts one million dollars in assorted personal debts.

John Mackay sold out his interest in the Comstock in 1883, convinced that there was nothing left but "poor man's pudding," and began an entirely new career. He organized the Commercial Cable Company and laid his own wire from New York to London and the continent. When he found that the monopolistic Western Union Telegraph Company would not relay his cable company's messages throughout the United States, he built up the Postal Telegraph Company as a competitive national network. Then, in 1901 he set out to lay the first transpacific cable to Manila, but in 1902, shortly before its completion, he died of a heart attack. He was 70, and had outlived the other Silver Kings by seven, 13 and 24 years respectively. His estate was valued at about $50 million, but only $305 of it consisted of property he had left behind in Virginia City, including an office safe, a desk, one eighth of a cord of wood and 18 empty burlap ore sacks.

Sometimes whole communities, not merely individuals, felt the exhilaration of striking it rich and were anxious to take their places among the cultural centers of the world. This civic pride expressed itself in the establishment of libraries, debating clubs and cultural societies and, most particularly, opera houses and theaters. The silver-mining town of Georgetown, Colorado, had a theater within two years of becoming a permanent town, and Piper's Opera House in Virginia City attracted the great Shakespearean actor, Edwin Booth, in a production of *Hamlet* so realistic that Ophelia's grave was actually dug into the bedrock beneath the stage by pick-wielding Cornish miners.

But perhaps the most ambitious display of civic pride was put on by Leadville, whose mines were to produce the handsome sum of $200 million, largely in silver, by the turn of the century. In the 1880s, when the boom was at its peak, Leadville had a population of 25,000 and possessed any number of uplifting institutions. There were seven churches—Baptist, Presbyterian, Methodist, Congregational, Episcopalian, Campbellite and Catholic—and two large hotels, the Grand and the Clarendon. The food at the Clarendon, famous throughout the West, was prepared by Mon-

A prodigious silver service for a miner's wife

In the late 1870s, a lady well known for expensive indulgence, Marie Louise Mackay, placed an order with the New York firm of Tiffany & Company. Mrs. Mackay wished an elaborate silver service for her splendid Paris mansion. Her husband, John Mackay, who was living austerely near his mines in Nevada, would supply the silver and defray all other expenses.

It took 200 Tiffany silversmiths two years to make the 1,214-piece collection. The larger pieces bore the recently commissioned Mackay coat of arms on one side, and Mrs. Mackay's initials on the other. The pieces included flatware and dishes, carafes, coolers, candelabra, lamps and cigar boxes.

By Mrs. Mackay's extravagant standards, it was a small purchase. Yet it cost her husband a fortune: 14,718 ounces of silver worth some $17,000, in addition to more than $100,000 for the renowned Tiffany craftsmanship.

While working on Mrs. Mackay's enormous order, Tiffany & Company keeps John Mackay posted on his account, noting receipt of another silver bar from his mines. The finished pieces filled eight chests.

CIGAR STAND

CANDELABRA FOR 29 CANDLES

CHAMPAGNE COOLER

LIQUOR SET

WATER CARAFE

sieur A. La Pierce, a chef who had been hired away from Delmonico's in New York.

Another noteworthy establishment was what the local press described as "an elegant art photograph shop" operated by a certain Mr. Needles. "The photographs decorating his parlor, the music from his organ, the brilliantly lighted room, all remind one of New York City. Miss Needles, talented sister of Mr. Needles, is assisting, and also paints in oil, crayon and water colors, natural as life." Nearby stood the shop of the Professor, who created works of art on human limbs and elsewhere. "Ladies come to me every day to have a monogram or some loved one's initials inserted into their skin. The habit prevails principally among the fashionable fancy women," asserted the Professor, "but also in some good circles of society which follow the fads of London and Paris." For only $200 he would tattoo a heraldic device or motto "on the upper arm, where it may never be detected, or on the knee, or the side of the thigh, wherever the fancy dictates."

Leadville also had the benefits of modern medicine, including those offered by Dr. Charles Broadbent, who sold Inhaling Balm for catarrh, Dandelion Pills for dyspepsia and biliousness, and a liquid called the Great English Remedy that would cure "Loss of Memory, Lassitude, Nocturnal Emissions, Noises in the Head, Dimness of Vision and Aversion to Society."

In addition, Leadville boasted a three-story theater in which the plays of Shakespeare and several operas were presented, the latter staged by the Emma Abbott English Grand Opera Company. A newspaper critic reported that Miss Abbott, rising above the dullness of the old librettos, "conceived and executed the idea of singing 'nearer My God to Thee' in the third act of *Faust,* introduced Siberian bloodhounds in *Lucia di Lammermoor,* interpolated 'Swanee River' in *King for a Day,* and had a trapeze performance in *Romeo and Juliet* and a trained mule in *Il Trovatore.*"

Despite the wealth and accomplishments of their city, the residents of Leadville as time went on felt that the rest of the world, particularly the metropolitan centers on the Eastern seaboard, were not fully aware of the recreational and cultural opportunities Leadville had to offer. Consequently, in the fall of 1895 they resolved to promote the city by means of a great Winter Carnival whose focal point would be an Ice Palace.

Other cities — Moscow, Montreal, St. Paul — had built ice palaces to celebrate winter carnivals, but Leadville's would be the biggest, most magnificent and most expensive edifice ever constructed of ice.

On November 1, 1895, a crew of 260 workmen that included 52 carpenters, working two shifts, began to erect wooden frames for the ice blocks, most of which were obtained from the lakes of the local water company. It was so cold in Leadville, at its elevation of 10,000 feet, that the ice was 20 inches thick. The blocks were hauled ashore with hooks and cables and dragged on sleds a mile and a half to the site of the palace. At first stonecutters were hired to trim the blocks to a standard 20-by-30-inch size, but when these craftsmen proved too slow, a number of Canadian woodsmen, equipped with broadaxes, were imported to do the job. The trimmed blocks were placed in the wooden frames and sprayed with water to freeze them into a solid mass. Work continued day and night for two months.

The completed palace resembled a huge, square Norman castle, 325 feet on a side, with five-foot-thick walls decorated with bands of projecting blocks, panels, corbels and buttresses. The main entrance was flanked by octagonal towers 90 feet high and 40 feet in diameter, with turrets and battlements. There were a half-dozen lesser towers at the corners of the palace and at the south entrance, all surmounted by tall flagpoles from which Old Glory strained stiffly in the cold mountain breeze.

Most visitors to the palace entered at the main gate on the north and ascended a spacious stairway of ice leading to an 80-by-190-foot skating rink. (Although the use of conventional materials was avoided wherever possible, this great room perforce had a wood-and-metal truss ceiling.) The rink was illuminated by electric lamps embedded in pillars of ice, from which light radiated in all directions. Nearby were the grand ballroom and a dining room, smaller than the rink but still impressive in size. These rooms were furnished with upholstered chairs and settees for the weary and were kept at moderately comfortable temperatures by large stoves.

On display everywhere were industrial, agricultural and mineral products of all sorts set inside great blocks of the glass-clear ice and so arranged that, as one writer

Marie Mackay paid $15,440 for her portrait (from which this sketch was made) by a French artist. But she hated it, so she burned it.

put it, "the palace seemed a vast art gallery." The T. G. Underhill Company showed frozen specimens of pants, overalls and vests; the Booth Packing Company displayed canned goods, oysters and fish in eight ice blocks; and the Denver & Rio Grande Railroad, in an exhibit of 15 blocks, presented a panorama of its route from Denver to Leadville. There were frozen shoes, frozen sewing machines, leather goods, soap, pickles and beer. Among the beer manufacturers, Adolph Coors of Golden, Colorado, took the precaution of filling his bottles with colored salt water, lest the bottles break upon freezing. This caused some distress among workmen who stole several bottles and presently found themselves gagging.

Throughout the palace there were statues made of snow, sprayed and frozen to a lustrous finish. At the main entrance stood a 19-foot ice sculpture not unlike the Statue of Liberty. But instead of holding aloft a torch, the lady was pointing her right arm and lifting up her eyes to the hills, whence had come Leadville's wealth. Draped over her left arm was a scroll bearing the gold insignia "$200,000,000."

Other statuary recounted a sort of Miner's Progress in four installments: a prospector and his burro; a hard-rock miner with drill and hammer; a triumphant miner at the moment of striking it rich; and a wealthy retired miner with a toothpick in his mouth and a silk top hat on his head, who had just sold his mine and was about to take a trip to Europe.

The palace was opened on January 1, 1896, and remained in good condition for a number of months. The south side, sheltered from the wind and exposed to the strong mountain sunlight, had to be protected with 10,000 square yards of canvas, but the rest of the structure held up well past the official Carnival closing of March 28. But any recognizable ice structure was melted and gone by mid-June.

Of all the tales associated with striking it rich on the mining frontier, perhaps the most instructive one—at least from the viewpoint of the bluenose moralist—was that of the rise and fall of H. A. W. Tabor, a sometime citizen of Leadville. Tabor, originally a Vermont stonecutter, married a Maine girl, Augusta Pierce, at the age of 27, and in 1857 the two of them went West to seek their fortune. Augusta was thin and angular, with a dour New England face that was made even more forbidding by a pair of pince-nez spectacles tinted a glacial blue. Yet Augusta was a woman of courage and loyalty, and had a keener eye than her husband for the beauty of the prairie and mountains.

For many years H. A. W. Tabor prospected with modest sucess in various diggings in Colorado. But it was Augusta who managed to keep the pair going. She made bread, butter and pies to sell, boarded miners in tents and log cabins that Tabor built, ran a general store and acted as unofficial postmistress and banker for miners, who trusted her and left their money and valuables with her when they went out to prospect.

By 1878, after nearly 20 years in Colorado, Tabor and Augusta were living in Leadville and Tabor was a fairly prosperous merchant and a respected civic figure. Customers were so numerous that he was often called on to give up poker, at which he was very good, and help his wife do the chores. One spring day two poor German prospectors named George Hook and August Rische came into the store and asked for a grubstake. They were not impressive to look at—one man who saw Rische at the time described him as "the worst played-out man I ever met, his entire wealth consisting of a pick and a spade and a faithful old dog." But they were persistent.

At length, simply to be rid of them, Tabor gave them $17 worth of supplies in return for a third of their earnings. Legend has it that they also helped themselves to a jug of whiskey, not part of their grubstake, and took off into the hills. After drinking the whiskey they dug a hole, choosing a shady site because they preferred not to work in the sun. The hole eventually became a 27-foot shaft, at which depth they struck the top of an almost-vertical vein of silver that ran straight down into the hillside. If they had dug only a few yards north or south, east or west, they would have missed it. As it was, the vein, called the Little Pittsburg, soon began to produce $20,000 a week, and Tabor, who owned one third of it because of his grubstake, sold out after a year for one million dollars in cash.

The mine then became part of a public corporation in which Tabor bought stock, quickly gaining another million as the shares went from $5 to $30. He began to pour his money into more than a dozen other mines—including Chrysolite, Matchless, Scooper, Union

Emma, Tam o' Shanter, Henrietta, Hibernia, May Queen, Elk, Little Willie and Wheel of Fortune — and grew still richer. The Republican Party in Colorado, anxious to put Tabor's wealth to public service, nominated and elected him in 1878 to a two-year term as lieutenant governor.

The hard-working Augusta was skeptical of such easy-won wealth and position. Her husband, now that he was required to spend a good deal of time in the capital, bought a large house in Denver, but she was reluctant to enter it. "Tabor," she said when she saw the house, "I will never go up those steps if you think I will ever have to go down them again."

Possibly she sensed the disaster that was coming. Tabor had begun to invest in fields he knew little about — real estate, a water company, a gas company, a horse-car company, a bank, insurance companies, lumbering companies and the Tabor Milling Company. He spent $1.2 million to gain control of a company that had been set up to build a great harbor and manufacturing center on Lake Michigan that would rival Chicago. He speculated in corn and thought of trying to corner the market in wheat.

He had also bought himself an immense diamond ring, much to Augusta's embarrassment, and he had joined the Bel Esprit Society of Leadville, listening to soprano solos and lectures on Cleopatra's Needle and the Seven Wonders of the World. He insisted that Augusta hire some servants, which she unwillingly did. In other regards she was less cooperative. She dressed plainly, refusing to wear the gaudy clothes he thought appropriate to a millionaire's wife, and when he held entertainments at his Denver house, she invited the servants into the parlor to hear the music.

In 1881, as his wealth approached nine million dollars and one of his mines, the Matchless, was alone providing him with at least $1,000 a day in pocket money, Tabor endeared himself to Denver by constructing the Tabor Grand Opera House. The building cost a total of $750,000, and was furnished with carved cherry brought from Japan, marble from Italy and heavy silk fabrics from France.

When the Opera House opened in September, the audience of 1,500 rose to its feet and cheered Tabor again and again. Although there were a good many who thought him a walrus-moustached bumpkin, and

who suspected that his financial empire was not very well glued together, no one wished Tabor ill. He was naïve, an innocent who hugely enjoyed operating in a sophisticated world of big money he did not understand. Although the Denver newspapers made fun of him, even some of his less friendly contemporaries were willing to say that he was "confiding, charitable, and generous. No man ever went to him for a favor when he had money and came away empty-handed."

Augusta Tabor was not present to hear her husband cheered when the Opera House was opened. The reason soon became evident when she filed a lawsuit against him, not for divorce but for support. He had deserted her more than a year earlier, and although his income — as her complaint said — "amounts to more than $100,000 per month in money" he had given her nothing, obliging her to support herself "by renting rooms in her place of abode and by keeping boarders." Tabor successfully maneuvered to have the suit set aside, but all Colorado was shocked. Indeed, all Colorado would have been a good deal more shocked if Tabor's recent affairs had been public knowledge. Two years earlier, at 50, he had befriended a beautiful 26-year-old divorcée named Baby Doe and now he intended to marry her.

"Baby Doe" was not a term of endearment but the name to which the young lady answered. Born in Oshkosh, Wisconsin, and christened Elizabeth McCourt, she married a ne'er-do-well named William Harvey Doe, and went West to Colorado with him in the 1870s. They settled in Central City where, according to legend, the appreciative miners referred to her as a "beautiful baby"; hence her name. After a few years of marriage she divorced Doe and journeyed for excitement to Leadville, where Tabor saw her and was overcome. As a newspaper described her, she was "without doubt the handsomest woman in Colorado. She is young, tall, and well-proportioned, with a complexion so clear that it reminds one of the rose blush mingling with the pure white lily; a great wealth of light brown hair, large, dreamy blue eyes, and a shoulder and bust that no Colorado Venus can compare with."

Early in 1883 Tabor was constantly in the news. He brought pressure to bear on Augusta and forced her to obtain a divorce, settling about $300,000 on her. When the hearing ended, the bewildered lady

Leadville's Ice Palace, with a 19-foot statue of Lady Leadville gracing its entrance, stands as a celebration of the town's silver wealth in 1896. Assembled at a cost of some $20,000, it required 5,000 tons of ice, covered five acres and remained open from New Year's until midspring, when warm weather made most of it too dangerous for occupancy.

turned to the judge and said, "What is my name?"

"Your name is Tabor, ma'am. Keep the name. It is yours by right."

"I will. It is mine till I die. Judge, I ought to thank you for what you have done, but I cannot. I am not thankful. But it was the only thing left for me to do. But, Judge, I wish you would put in the record, *Not willingly asked for.*"

Augusta walked out of the court in tears, repeating, "Oh, God! Not willingly, not willingly!"

While the divorce hearing was in progress the Colorado legislature met in Denver to elect two United States senators, one for a full term and one to fill a vacancy for only 30 days. Tabor was hopeful of getting the long term, having distributed about $200,000 among various Republican Party officials, but apparently his divorce weighed against him. He lost on the 97th ballot and had to accept the 30-day term.

During his one month in Washington Tabor created quite a stir. When he was sworn in, he wore a large diamond ring on either hand, and huge cuff links of diamonds and onyx in a checkerboard pattern, causing one fellow senator to murmur "Grrreat God!" A few weeks later he married Baby Doe in a dazzling ceremony at the Willard Hotel for which he bought her a $7,000 wedding dress and a $90,000 diamond necklace. President Chester A. Arthur himself held her hand, saying, "I have never seen a more beautiful bride."

After they returned to Denver, Tabor made a second unsuccessful try in 1886 for the Senatorial nomination, and in 1888 he failed in a bid for the Republican nomination for governor. The party higher-ups apparently still felt that his marital adventures would not sit well with the voters. Meanwhile he lavished immense sums upon his young wife. He bought her an ornate Italian-style villa and presented her with several regal carriages in which she enjoyed driving around the city; her favorite was enameled in dark blue with gold stripes, lined with pale blue satin, and drawn by four black horses. He gave her jewels by the handful and commissioned five oil portraits of her, each in a different costume and pose. In the first few years of marriage two daughters were born. One was called Elizabeth Bonduel Lillie, but by the time the second arrived the Tabors had the hang of naming children and she

was christened Rose Mary Echo Silver Dollar. However she would see only a brief period of prosperity. Her father's fortunes were beginning to fall apart.

Tabor's financial collapse had been coming for a long time. In the mid-1880s one after another of his businesses and mines had petered out, and in his attempts to develop new ones he mortgaged all his real estate. The financial panic of 1893, during which 12 reputable banks in Denver were forced to close in one three-day period, nearly wiped him out. The Matchless mine still produced a trickle of silver but even that lost much of its value when the Sherman Silver Purchase Act was repealed, also in 1893, and the government ceased buying 4.5 million ounces of silver a month for coinage.

Baffled and desperate, Tabor speculated in gold mines with the last of his capital and lost it. Baby Doe's jewelry and his own, along with his villa, his carriages and his stables were sold at auction. (No doubt Augusta, living in seclusion since the divorce, read of this. She died in 1895.) In 1897, nearing 70 and by now flat broke, Tabor went out alone to prospect vainly in the hills. After a year of hardship he was rescued by Colorado's two senators, who got him a job as postmaster of Denver. The salary was barely enough to allow Tabor, Baby Doe and their two daughters to live in one small room in a hotel, where he died in 1899. Among his last words to his wife were, "Hang on to the Matchless," his one remaining property, which he thought might someday make her rich again.

Tabor's story was neatly rounded: rags to riches to rags, his life closing with the century and with the era of gold and silver in the West. But the story of Baby Doe had no such pat, timely conclusion. Although she was a woman of the 19th Century and although her life in effect ended with Tabor's, she neglected to die at the appropriate moment. She went back to Leadville and moved into a dilapidated tool house near the shaft head of the Matchless. Year after year she lived there, hanging on to the mine as Tabor had instructed her to do. No one wanted the mine, which was exhausted and full of water, but she held on for a very long time, until almost everyone had forgotten her. It came as an eerie shock to historians when the frozen body of the 80-year-old woman was found, dead for a week or two, on the floor of the shack in 1935.

Silver Baron Horace Tabor *(top)* created a sensation in 1883 by divorcing his stern-faced wife Augusta *(above)* in order to marry a beauty named Baby Doe *(in ermine cloak)*. President Chester Arthur was among the dignitaries at their wedding in Washington, and Tabor had dreams of a grand political career. But all turned to ashes when Republican Party elders scorned him and the silver in his mines ran out.

The social ascent of an unsinkable lady

By and large, the consorts of the gold and silver kings were remarkable mainly for their conspicuous consumption. But there was one lady who came to a good bit more. She was Mrs. James J. Brown, whose husband made his fortune in Colorado gold.

Born Maggie Tobin, in 1867 in Hannibal, Missouri, she started life as a ditch-digger's daughter. Pretty and vivacious, she married Jim Brown at 19; he was a mine manager in Leadville, and soon was able to give her all she had ever hoped for: clapboard house, steady income and two children.

Life grew complicated—and wonderful—in 1894 when Jim struck a vein of gold and swept her off to an 18-room mansion in Denver. Carved Nubian slaves decorated the foyer, while the parlor exhibited ornately framed paintings and a rare pianoforte. With such a showplace, Maggie felt ready to take her place in society. She posted invitations and made preparations for gala soirees. But Denver's upper crust wanted no part of the nouveau riche and obviously lower-class Browns. Maggie had to call in neighborhood kids to eat the feasts her chefs had prepared.

However, she was nothing if not determined. To correct her lack of education and polish, she hired tutors, read all the best books and toured Europe several times. In time, she became fluent in several languages, was a witty raconteur and grew to be friends with the Astors, Whitneys and Vanderbilts she had met on her travels.

Maggie—who by now had adopted the more voguish diminutive Molly—did not conquer Denver until 1912. At that time she won the entire nation's admiration for her heroism on the sinking liner *Titanic*. Taking command of a lifeboat, she organized the rowers and nursed the injured. When reporters later asked Molly how she had survived, she lightly replied, "I'm unsinkable." And thus the Unsinkable Molly Brown became a part of American lore.

Foliage bedecks the foyer of Molly Brown's mansion. She is shown in the inset with husband and children.

180

A large collection of leather-bound books — she actually read them — lines Molly's library.

Outstanding in Molly's dining room are two tapestries, flanking the china press at left.

A gigantic polar bear rug dominates Molly and Jim Brown's flower-filled parlor.

6 | Long trail to the Klondike

"I never saw men work harder, bear more and accomplish less," a prospective miner named Will Langille wrote in a letter back home to Oregon. And Jack London afterward recalled that "every back had become a pack saddle and legs became drunken with weariness." Both men were talking about one of history's biggest, wildest and, in the end, most futile gold rushes—the incredible stampede to the Klondike in Canada's forsaken Yukon.

Nobody who climbed the Chilkoot Pass on the trail north from the Alaskan panhandle in the winter of 1897-1898 (below) ever forgot the agony

of that wrenching toil. But whether the Klondikers scaled the Chilkoot or chose nearby White Pass or took the long way up the Yukon River, there was scarcely an easy mile on any of the routes to the gold fields.

On the Chilkoot, avalanches and glacial fissures were a constant threat. On White Pass, the trails were barely two feet wide in places—a broad highway for mountain goats but a potential deathtrap to scores of overburdened animals and snow-blinded men. On the Yukon River, the raging waters in Miles Canyon and the rapids beyond drowned 10 men before the North-

West Mounted Police imposed stringent safety regulations, and overhanging trees called "sweepers" knocked many other voyagers overboard.

Thousands of men—and more than a few women—finally reached their Eldorado in Dawson in the heart of the Yukon. But nobody ever made a more harrowing pilgrimage for less. Getting there with their food and equipment had cost them somewhere between $30 and $60 million. The Klondike's total yield in 1898 was only around $10 million. All told, only one out of 10 who reached their Dawson City goal in 1898 found any gold at all.

Klondikers take "icy steps to hell," as one of them described the ascent of the Chilkoot. The tower at top is part of an early aerial tramway.

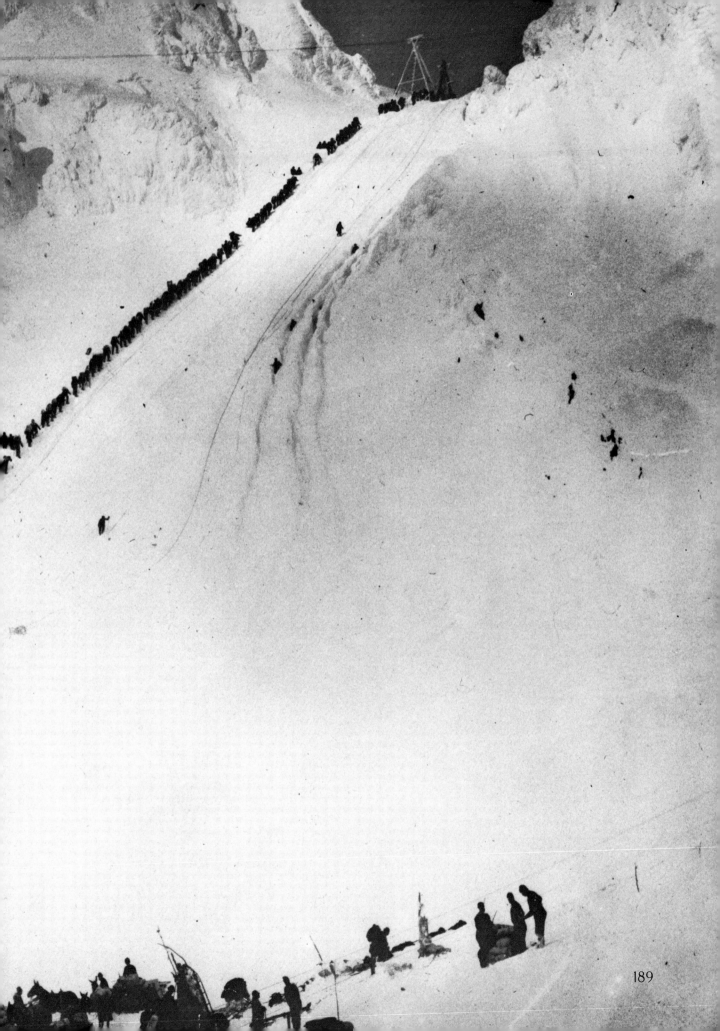

Beached on the tidal flats at Skagway Bay, a scow full of supplies for the Klondike is off-loaded into horse-drawn wagons. The bay was so shallow that ocean-going vessels had to anchor a mile from shore. The long rows of piles in the background were the beginnings of deepwater wharves.

191

Clamped between granite walls, Klondike-bound trekkers rest in "Canyon City" — a camping ground nine miles uphill from their landfall. Horses and wagons could barely make it through the boulder-strewn gorge and were invariably abandoned at Sheep Camp, a few miles before Chilkoot Pass.

Three Klondikers in a dory shoot the Yukon's White Horse Rapids. "Grazed death by barely half a dozen inches," said a survivor as he reported the fate of a less fortunate party: "one-two-three—each of Major Walsh's three boats reared high in the sleety mist and overturned one after the other."

195

Impatient Klondikers wait out one of many traffic jams on the Porcupine Hill sector of the White Pass trail. The path was narrow, crooked, often knee-deep in the mud of melting snow, and so exasperating that it summoned forth profanity that one traveler described as "sublime in its awfulness."

THE KLONDIKE NEWS

VOL. I.   DAWSON, N.W.T. APRIL 1ST, 1898.   NO. I

OUTPUT FOR 1898 $40,000,000.

KLONDIKE NEWS

FROM N° 8 EL DORADO,
PROPERTY OF CHAS. LAMB,
VALUE $315 00

SCHMIDT L. & LITH. CO. S.F.

DISCOVERER,
GEO. W. CARMACK.

THE LARGEST GOLD NUGGET,
FOUND IN EL DORADO CREEK NO.36 BY M. KNUSTON
WEIGHT 26 OUNCES VALUE $ 530 00

A cruel adventure for 100,000 optimists

One evening as the long winter of 1897-1898 closed in on the Yukon, a 22-year-old French-Canadian girl named Mabel LaRose was boosted up to stand on top of the bar in the Monte Carlo Dance Hall and Saloon in Dawson City. She was wearing her best dress; her auburn hair was neatly braided; and she took a serene if slightly wistful disregard of the remarks of the miners who had gathered to examine her. Although there were prostitutes in Dawson City, she was not, exactly, of their number. She was a dance-hall girl who was paid to flirt outrageously with the patrons but who, as she put it, "lived private."

Recently, however, Miss LaRose had modified her professional status by making a bold offer: for the duration of the winter she would serve as wife in every respect to the highest bidder. There were a few stipulations. The money was to be held by a neutral party until the completion of the bargain; temporary husband and wife were to treat each other kindly; and if Miss LaRose did not like the looks of the highest bidder, she had the right to reject him. For several minutes she stood on the bar while the miners made their appraisals. A flurry of bidding followed and, as at a livestock auction, Miss LaRose was knocked down for $5,000 in gold.

In a similar public sale of a dance-hall girl in Dawson shortly thereafter, the purchaser proposed a more permanent union: he offered to marry the young lady in question. At first, she refused—not because she disliked his appearance but because she knew that he was drunk much of the time, sobering up only long enough

to work his rich mining claim. This decision met hoots, cheers and a babble of shouted advice from the onlookers. At length the miner offered the girl her weight in gold if she would change her mind. Although she was a mere slip of a thing and had no time to fatten up, she accepted and was weighed forthwith. She checked in at 112 pounds, which, at $14 an ounce, brought her reverse dowry to $25,000.

It could only have happened in Dawson City, a metropolis that, during the winter of 1897-1898, was fully as uncommon a place as some of the marital compacts negotiated there. For one thing, it was an instant city, being only a year old in the season of Miss LaRose's betrothal. It was also unquestionably one of the hardest places in North America to reach. It lay in the Klondike region of the Canadian Yukon, at the heart of a wilderness of tundra and forest that spread across an area almost the size of Texas. There were only two practical ways to reach Dawson. One was from the west, 1,700 tortuous miles up the treacherous Yukon River, which was frozen solid much of the year. The second approach was from the south, 360 agonizing miles from the coast of Alaska through an all-but-impassable barrier of mountains, glaciers and rapids.

Dawson City was so inaccessible, in fact, that it was a wonder Miss LaRose had the gumption to get there at all—though no wonder at her value once she did. For, if women were scarce in the Klondike, men were present in astonishing numbers. From 1897 to 1899, tens of thousands of single-minded males made the incredible journey to Dawson in what was the last, the shortest and surely one of the most bizarre and exuberant gold rushes of all time—a rush that would be remembered longer for the travail and the triumph of getting there than for the treasure that lay at its end.

The breed of men who had found gold in California and then spent half a century chasing it back and forth

---

Fate came to George Carmack, the Klondike's first claimant, in a memorial newssheet dated 1898, but printed in 1909. The headline was a wild exaggeration.

Barricades of provisions and equipment line
Seattle's sidewalks as miners await transportation to the docks and passage thence
to Alaska. Businesses like Cooper & Levy
reaped a $25 million bonanza in 1897.

across the spine of the western American mountains
were not the sort to be discouraged by either terrain or
climate. Their lust was such, as one veteran recalled,
that "the only thing needed to reach the North Pole
is assurance that there is gold . . . then in less than
six months there will be a highway cut, and the
pole changed to a May Pole with the American flag
waving on top."

Prospectors first reached the Yukon Territory
around 1870, and soon half a dozen grizzled veterans
were scratching at creek beds draining into the Yukon
River. By the 1890s there were perhaps 2,500 prospectors in the Klondike region, mostly working near
two crude log towns on the Yukon: Circle City in
American territory and Forty Mile (so named for its
distance from the region's main trading post) on the Canadian side of the border. In 1895 they panned more

than one million dollars worth of bright yellow metal
—a handsome enough total but, divided among them,
not enough to make many of them rich.

Nor was it enough to impress many people down
south in California, Oregon and Washington. Despite
all the fabulous gold and silver strikes throughout the
West, people had developed a certain pained skepticism about tales of new finds. Too many men had
gone stampeding into the hills chasing rumors that led
to nothing. If the storytellers of the Yukon wanted anyone to believe them, let them send down some real gold
—and not a mere handful, either. It would take a ton of
the stuff to be convincing.

As it happened, the miners were about to do better
than that: two or three tons. In the summer of 1896
they were investigating one of the Yukon's tributaries,
a little river called the Thron-Diuck by local Tagish In-

200

dians. The miners had difficulty pronouncing it since the Tagishes spoke with a guttural gargle, but did the best they could, calling it the Klondike.

In mid-August on one of the small branches of the Klondike, a Canadian named Robert Henderson made a strike that yielded eight cents a pan—nothing remarkable, but still a good prospect. Henderson named the place "Gold Bottom Creek" because, as he said, "I had a daydream that when I got down to bedrock it might be like the streets of the New Jerusalem." Unfortunately Henderson's strike did not turn out to be worth much, although it won him a small reward. The Canadian government in later years recognized him as one of the codiscoverers of the Klondike's major gold field, and gave him a lifetime pension of $200 a month. The other man credited as codiscoverer was an American, George Washington Carmack, who got

so much gold that he needed no pension from anyone.

Carmack was not really a prospector; he was a squaw man who wanted nothing more than to be an Indian. Born near San Francisco, the son of a forty-niner, Carmack had made his way to the Yukon as a youth and married a Tagish Indian chief's daughter. He lived lazily among his in-laws, hunting and fishing, and hoped someday to succeed to the chieftainship.

Not long after Henderson made his strike, Carmack and two Indian friends, Skookum Jim and Tagish Charlie, came on a far richer deposit about 10 miles away on Rabbit Creek, another branch of the Klondike. In his first pan Carmack found a quarter of an ounce of gold, worth about four dollars, and in a few minutes he secured enough dust to fill an empty shotgun shell. He blazed a spruce tree beside Rabbit Creek and penciled a claim notice on the bare wood. As the discoverer of the placer, he took two claims while the Indians got one apiece. Each claim was 500 feet long and extended from bank top to bank top.

Carmack wasted no time in setting out for the town of Forty Mile to record the claims at a Canadian North-West Mounted Police post there. On the way he encountered half a dozen wandering prospectors and told them of his discovery. In Forty Mile Carmack continued to talk proudly of his strike, pouring the gold out of his shotgun shell in a saloon for all to examine. Almost at once, as though by some occult telegraphy, the news began to spread up and down the Yukon valley and prospectors headed for the creeks of the Klondike River. Although the gold was in Canadian territory, most of the men who went after it were Americans. Within a few weeks they had staked dozens of 500-foot claims on Rabbit Creek—hastily renamed Bonanza Creek. The miners then turned to one of Bonanza's promising branches, Eldorado Creek, staked it from start to end, and began to dig.

As shortly became apparent, the area of gold distribution was only about 25 by 30 miles in size and had about 300 miles of creeks. At 500 feet each, there was room for fewer than 4,000 claims—and most of these proved unproductive. There was also room for a large number of hillside or bench claims, which were limited to squares 100 feet on a side. With great labor, a man could sink a shaft in a hill, and with even greater good luck he might hit the bed of an ancient stream and find gold. But that was most uncommon; the usual result was a "skunk," or barren hole.

Even for the fortunate few, the Klondike held hardships. This was a country of low hills covered with white birch, cottonwood and spruce. Small game was plentiful, as were caribou and moose. The rivers and creeks were full of trout and salmon. But when winter came, a long season that clamped the land in twilight and dark, the cold was crushing. The miners had no thermometers that could keep accurate track of it, but they improvised by filling vials with liquids whose freezing points were known. When mercury turned solid, it was −38°F., a comparatively balmy temperature fit for hunting, traveling by dog sled or any outdoor work. When good strong whiskey froze, it was −55° and chilly. Kerosene solidified somewhere around −65°, and when a ferociously alcoholic medication

called Perry Davis' Pain-Killer went hard, it was −75° and damned cold.

Life in the miner's cabins was not cozy. Made of green tree trunks chinked with moss, the cabins were about 12 by 12 in size, low-roofed, with small windows made of empty glass jars or bottles held together by hard mud. The indoor temperature was often far below freezing. Steam from cooking formed huge icicles or "glaciers" on the cabin walls, from which ice was chopped to be melted for more cooking and drinking. The ax used for splitting kindling had to be kept under the oven lest the cold make the metal so brittle that it would shatter on impact with the wood.

The brief summers, with more than 20 hours of light a day, were surprisingly warm. Sometimes the temperature rose to 100°, and with the warmth came an affliction that men seriously compared to the plagues of Egypt: mosquitoes. "Apparently no larger than the ordinary mosquito of lower latitudes, they are several times as venomous," wrote an appalled observer. "One may hurl a blanket through a cloud of them, but ranks are closed up and the cloud is again intact before the blanket has hit the ground. They rise in vast clouds from the peculiar moss along the banks and creeks, and their rapaciousness knows no limits. They have been known to drive men to suicide, and the sting of a few dozen will make a man miserable for days. I have seen tough miners sit and cry."

The "peculiar moss" in which the mosquitoes bred was sphagnum, or muskeg. Across much of the north it covered the ground to a depth of more than a foot; when it thawed in summer it provided an ideal en-

---

TREACHEROUS ROUTES TO DAWSON CITY

The most popular route to the Klondike was a 550-mile trek north from the Skagway or Dyea ports in the Alaskan panhandle, through an obstacle course that seemed designed by the devil himself. On the first leg (inset), over the mountains to Lake Bennett, there was a choice: from Skagway through the White Pass with its alternating precipices and fearful swamps, or from nearby Dyea via the even more precipitous but thankfully swampless Chilkoot Pass. Once over the passes, the argonauts had to negotiate a series of wind-whipped lakes and rapid-strewn rivers before finally arriving at Dawson. Another route (not shown) was by steamer around the Aleutians to St. Michael and thence 1,355 miles up the Yukon. It was relatively safe, but could take many months, and few Klondikers were inclined to wait.

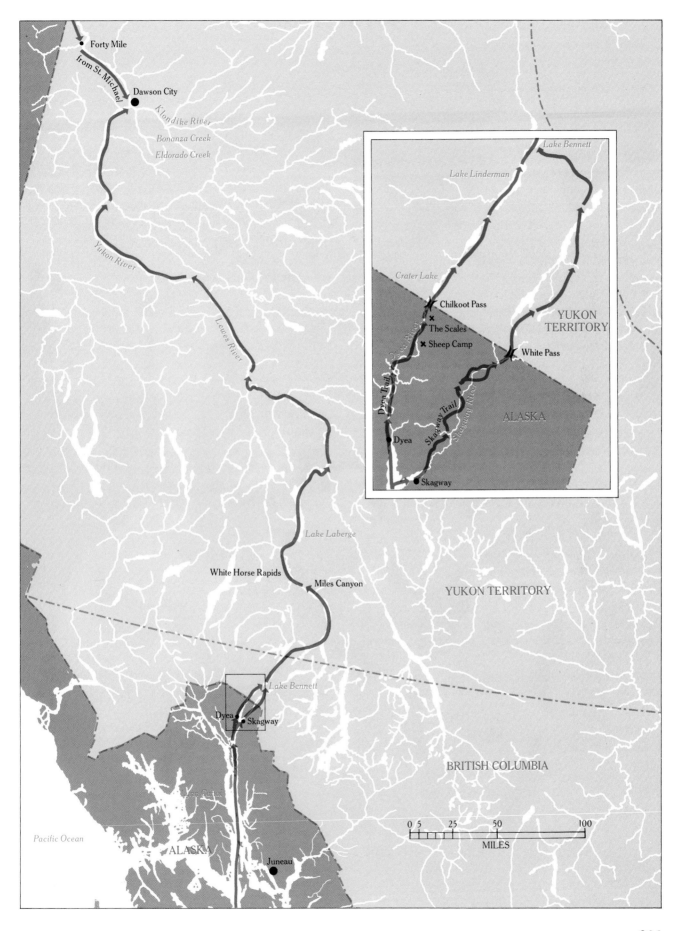

Forty Mile

Irom St. Michael

Dawson City

Klondike River

Bonanza Creek

Eldorado Creek

Yukon River

Lewes River

Lake Bennett

Lake Linderman

Crater Lake

Chilkoot Pass

✗ The Scales

✗ Sheep Camp

Dyea River

Dyea Trail

Skagway Trail

Skaguay River

White Pass

YUKON
TERRITORY

ALASKA

Dyea

Skagway

Lake Laberge

White Horse Rapids

Miles Canyon

YUKON TERRITORY

Lake Bennett

Dyea
Skagway

BRITISH COLUMBIA

Lynn Canal

Pacific Ocean

ALASKA

Juneau

0 5   25   50   100

MILES

Rescuers exhume a victim of a Chilkoot avalanche on April 3, 1898. Although warned of the danger created by a two-week blizzard, Klondikers were on the pass in force when the slide came roaring down. Of some 70 climbers buried under 30 feet of snow, only a few were dug out alive.

vironment for the insect larvae. Beneath the muskeg lay a deep stratum of permafrost that never thawed at all. And it was this rock-hard permafrost that the Klondike miners had to penetrate to reach their gold. True, dust and small nuggets could be panned from the creek beds, but the real treasure lay on the bedrock below, from 10 to 50 feet down through the permafrost.

Pioneers in the Klondike region at first tried to reach bedrock using only the heat of the sun—every day they scraped away an inch or two of subsoil where the permafrost had been softened, and repeated the process all summer. But by the time George Washington Carmack made his discovery, this inadequate method had been replaced by what they called fire setting. At nightfall the miners covered the bottoms of their shafts with charges of wood that would burn to ashes by morning. The heat would soften up to six inches of subsoil. So the bedrock with its overlying pay streak might be reached in only a month or two. Lateral drifts could then be melted along the streak, requiring no timbers because the frozen earth supported itself.

The fire-setting method could be used throughout the winter, although in that case the miners could not be certain of the extent of their good fortune or lack of it. The subsoil immediately refroze when it was hoisted to the surface and there was no way of washing it until spring. Frequent tests were made indoors in buckets of melted snow, but almost all the dirt was piled in large dumps near the shaft mouths, to be washed in sluices when the creeks gushed with water in the spring.

From late summer 1896 through the following winter, in diggings that looked like so many miniature volcanoes, men worked Bonanza and Eldorado creeks. At the confluence of the Klondike and the Yukon rivers a town sprang up and men made their way there out of the wilderness. The town was named for George Mercer Dawson, chief of the Geological Survey of Canada, and was soon to have a population of 25,000; before long, real estate would sell at $5,000 a front foot on the main street. But as 1896 ended, only rumors of gold had reached the world to the south.

It was not until July 1897, after the winter's diggings had been washed and several dozen miners had brought their gold down to the American West Coast that the great news broke. The miners reached the mouth of the Yukon on the Bering Sea in little stern-wheel steamers that plied the river in summer, catering to the fur trade. Thence they sailed south in ocean-going vessels, the *Portland* for Seattle and the *Excelsior* for San Francisco, carrying gold in tin cans, pickle jars, suitcases, blankets and caribou-hide bags.

GOLD! GOLD! GOLD GOLD!
Sixty-Eight Rich Men on
The Steamer *Portland*
STACKS OF YELLOW METAL!

The headlines in the Seattle *Post-Intelligencer* were echoed around the world. Crowds gathered at the wharf to watch the miners come ashore and were popeyed at what they saw: a onetime coal miner named John Wilkinson trying to lug $50,000 worth of gold in a small suitcase so heavy the handle snapped off; another man struggling with a leather poke, not much bigger than a grapefruit, that weighed 100 pounds; a man named Nils Anderson dragging a weighty sack down the gangplank and giving it to his wife, who had no idea he was rich until he told her he had brought out $100,000.

William Stanley was a lame, gray-haired bookseller from Seattle. He had a wife, seven children and a lot of debts. In a last effort to support his family he had gone north, taking one son with him, and now he was returning with $112,000. When she heard the news Stanley's wife, who had been working as a washerwoman, merrily walked away from her tub and told the customers to fish out their own clothes.

George Washington Carmack arrived in Seattle with his Indian wife Kate, Skookum Jim and Tagish Charlie. Among them, they had hundreds of thousands of dollars. They took rooms at the Seattle Hotel and Tagish Charlie caused a near riot by throwing 50-cent pieces (in those days enough for a lavish steak dinner) out of the windows to crowds in the street. According to one account Kate found the hotel's stairs and corridors a bewildering labyrinth and blazed a trail from the front door to her room by gouging chips out of the bannisters and doorframes with a knife. She had no difficulty paying for the damage.

The rush to the Klondike began almost instantly, at fever pitch, and for a half-dozen reasons. Telephones and high-speed newspaper presses had become available to spread the word far and fast. The Klondike diggings were placers—poor men's mines—where a res-

olute lad needed only ambition and luck to make his fortune. And there was no shortage of poor men in the country in the summer of 1897. A depression, caused in large part by a lack of gold bullion to back the nation's currency, had been going on since 1893, striking with particular severity in the Pacific Northwest. In Washington state thousands of people were living on clams dug from the flats of Puget Sound, consuming so many that, as a Tacoma congressman said, "Their stomachs rose and fell with the tide."

Altogether about 100,000 people headed for the Klondike, most of them from the United States, although many came from as far away as Australia and South Africa. But the 100,000 represented a far greater constituency. In hundreds of stores and factories workers pooled resources, selected a particularly vigorous or resourceful individual and financed his journey, hoping that he would make them all rich. Families held council; Uncle Jed or brother John would go while the others backed him. Farm groups, church groups and social groups sent representatives. In the Sacramento valley of California, fruit fell to the ground and rotted because the pickers took off. Half the recently graduated physicians in the state departed. The Boston newspapers reported that "Fourteen Hebrew working men of this city out of employment are determined to walk to Alaska or lose their lives in the attempt." Everywhere there were urgent want ads: "For sale. $400 buys sewing-machine store. Leaving city. Good chance for smart party." "Fine saloon in lively neighborhood. Place worth fully $1,500 but offered cheap as owner will leave city soon."

The most practical route to the gold fields was the way the miners had come out—by ship to the mouth of the Yukon and thence by riverboat to Dawson. But the riverboats were so few and small, and so immobilized by ice most of the year, that not many Klondikers managed to make the trip on them. The majority sailed to Dyea or Skagway, flimsy tent and log towns at the end of a long fjord in the Alaskan panhandle, debarked there, and went on foot over the Alaskan coastal range, using the White Pass or the Chilkoot. On the other side they built boats and drifted, sailed and paddled more than 500 miles north to Dawson on a chain of lakes and rivers. On the map the journey appears a fairly simple one, but to the people who actually completed it, it was the paramount experience of their lives.

As the rush began, ship fares from Seattle to Alaska rose from $200 to $1,000. All the available vessels on the Pacific coast, including many that were hopelessly leaky and unseaworthy, were rushed into service. So many people clamored for northbound berths that 18 transatlantic passenger liners steamed full-speed around Cape Horn to accommodate them.

The demand for baggage space was equally great. Each of the gold hunters took—or was urgently advised to take—about 2,000 pounds of supplies, checking off items from lists provided by newspapers, merchants and publishers of instant guidebooks. A typical outfit included 500 pounds of flour, 200 of bacon, 100 each of sugar, beans, dried fruit and dehydrated potatoes, plus another 500 pounds of assorted foodstuffs. Most men also carried the equipment required to build a boat and a cabin and operate a mine—gold pans, pick, shovel and spade, hammer and nails, whipsaw, brace and bits, chisels, calking irons, plane, pitch, oakum, rope and duplicate handles for all tools. Beyond that, there were cooking utensils, stove, heavy clothing, tent, gun and ammunition, soap, candles, fishing tackle, lantern, blankets or sleeping bag, medicines, tobacco, whiskey and a book or two. The merchants of Seattle, the principal port of embarkation, did $25 million worth of business in a few months.

Since guidebooks pointed out that pack horses and dog teams were useful in the Klondike, gold hunters bought large numbers of animals. Many were wornout, bony nags hurriedly shipped in from Montana, worth $3 one week and $50 the next. Miners also bought burros, sheep, goats and even reindeer with amputated horns, said to make excellent beasts of burden.

As to dogs, the Klondikers pounced on any creature that weighed 40 pounds or more. Seattle householders had to keep their pets locked up for, in this emergency, the rushers did not consider dognapping dishonorable. It was only one of the Klondikers' accumulating disillusionments to learn, in Skagway, that a domesticated state-side mutt was no Malemute and most likely never would be. Soon the streets of that town swarmed with abandoned and starving curs. It must be noted that the town's arch criminal, Soapy Smith *(page 147)*, had at least one small good side to his nature: he

adopted six strays personally and launched an Adopt-a-Dog campaign among his fellows.

Entrepreneurs both honorable and shady offered schemes, machines and gadgets. There were "ice bicycles" with stubby skis for front wheels, X-ray machines to locate gold, and hot-air balloons guaranteed to soar over the mountains and drop men comfortably in the gold fields. Dr. Armand Ravol, city bacteriologist of St. Louis, made plans to wipe out the mosquitoes by releasing germs deadly to the insects. And an outfit called the Trans-Alaskan Gopher Company offered shares at one dollar apiece to breed and train the industrious animals to dig tunnels in frozen ground.

If some of the schemes seemed more than usually lunatic, perhaps it was because the entire Klondike adventure was slightly mad. Most of the participants were city folk, sedentary office workers who had only the dimmest notion of the hardships that awaited them. They paid small heed when an Ottawa newspaper advised them to ask themselves, "Am I physically sound in every way and able to walk thirty miles a day with a fifty-pound pack on my back?" On August 10, 1897, the American Secretary of the Interior, C. N. Bliss, issued a warning against anyone trying to reach the Klondike that year. But even while Bliss was making his appeal another 2,800 people sailed north from Seattle, and hundreds more sallied forth from Tacoma, Portland, San Francisco and Victoria.

Meanwhile, because the Klondike lay in Canada, the government in Ottawa braced itself for the rush. Previously the only law enforcement in the deep Northwest had been provided by 19 constables of the Royal Canadian Mounted Police at Fort Constantine on the Yukon near the Alaskan border. Now another 80 men were added; before the rush ended there would be 285 Mounties in the area. Because jails were scarce—and insecure—the Mounties' favorite form of punishment for minor offenses was to sentence the miscreant to a period of chopping kindling on the town woodpile.

By the autumn of 1897, close to 35,000 people had reached the Alaskan coast at Dyea or Skagway. These twin towns of planks and canvas had no natural harbors; ships were anchored offshore while passengers and cargo were ferried in scows and canoes across the mud flats at high tide. Horses, dogs, sheep and other livestock were thrown, terrified, into the water and

Resilient womenfolk of the arctic gold fields

"What advice would I give to a woman about going to Alaska? Why, to stay away, of course. It's no place for a woman." Arriving in the Klondike as a brand-new bride in the fall of 1896, Mrs. Clarence Berry had to hike 19 miles with her husband from Dawson to their gold claim. "When I got there, the house had no door, windows or floor, and I had to stand outside until a hole was cut for me to get through." There were no facilities for bathing —save in the pan they used for washing gold. And when she ventured out on the trail, she was "covered with mud to the waist."

But she stuck it out and when the Berrys returned to California, eight months later, they were millionaires.

Only about 100 women reached the Klondike over the passes in 1897, but thousands more made the journey in 1898 and 1899 as the conditions improved: actresses, dance-hall girls, washerwomen, cooks and increasing numbers of wives following their menfolk. Hardly anyone came away as wealthy as Mrs. Berry, but many of the ladies did well, and some displayed a powerful entrepreneurial talent.

Belinda Mulrooney opened a prosperous roadhouse and was able to buy shares in six lucrative mines. Harriet Pullen, a 37-year-old widow, reached Skagway with hardly a dime. She earned enough selling pies to start a freighting business over White Pass, and eventually parlay that into ownership of a thriving hotel.

By all accounts, the grandest lady of the Klondike was Mrs. Eli Gage, a Boston-bred beauty whose husband was an auditor for a trading company. It was a rough life, she agreed, but a fascinating one. She found the miners to be "the kindest, most considerate and most practically honest people that I ever met"—qualities she doubted that she would find to the same degree "in Chicago or other civilized places."

Skirts hiked and satchels in hand, four rubber-booted actresses ford the Dyea River, while a fifth is carried across. The last woman in line still wears the mosquito netting that was essential protection against swarms of insects encountered on the trail.

The proprietor of Mrs. Lowe's Dawson City laundry stands in the doorway of her establishment. The lady not only laundered the shirts and underwear hanging on the line to her left; for a fee, she would predict where a raiment's wearer might find gold.

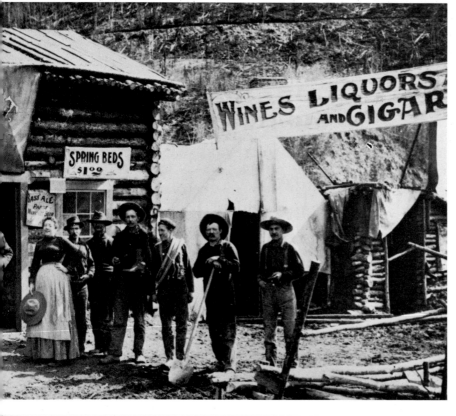

The Magnet roadhouse, owned by Belinda Mulrooney (believed to be fourth from left) stood near the mines at Grand Forks, 15 miles from Dawson. The place was as good as its name and brought her a small fortune.

forced to swim ashore. On the beach there quickly accumulated a tottering rampart of boxes and rope-tied canvas parcels, bursting open, water-soaked and splattered with mud, while cursing men waded through the ooze trying to locate their property and drag it inland.

In these circumstances it forcibly struck the Klondikers that no man could go forward or even survive alone. They formed partnerships, often after only a few minutes' talk, and bound themselves with a handshake. One of the immediate advantages of partnership was that the weight of baggage was lessened. The amount of food remained the same, but it was not necessary to carry duplicate sets of tools and utensils—unless the possibility of a future split-up was borne in mind.

Having organized and stripped down their possessions to what seemed a safe minimum, the Klondikers started inland toward the headwaters of the Yukon River. Those who had horses, mules or oxen, or could pay a packer to transport their goods, took the 45-mile Skagway Trail across White Pass to Lake Bennett. It started as an easy wagon road but turned into a narrow, nightmarish track that wound through quagmires and canyons, across streams, boulders, rivers and mountains, and along edges of cliffs where a misstep would drop a man or a horse 500 feet to the rocks below. A small-town boy from Plymouth, Indiana, wrote home about it: "I have been working like a slave since I came here trying to get over the trail and am not over it yet, and furthermore do not think I will be in time to get to the Yukon this winter. Since I came in we have lost our mule and one horse. I am undoubtedly a crazy fool for being here in this God-forsaken country but I have the consolation of seeing thousands of other men in all stages of life, rich and poor, wise and foolish, here in the same plight as I."

While the situation was bad enough for men, it was ghastly for the animals. Horses were of no value at the far end of the trail, where the journey was continued by water, and many Klondikers believed that if a beast could be kicked and flogged to the shore of Lake Bennett, that was sufficient; thereafter it could drop dead.

A young man named Jack London, who had not yet become a writer but was struggling along among the Klondikers with volumes of Milton and Darwin in his pack, noted the scene and never forgot it: "The horses died like mosquitoes in the first frost and from Skagway to Bennett they rotted in heaps. They died at the rocks and they starved at the lakes; they fell off the trail, what there was of it; in the river they drowned under their loads or were smashed to pieces against the boulders; they snapped their legs in the crevices and broke their backs falling backward with their packs; in the sloughs they sank from sight or smothered in the slime. Men shot them, worked them to death and when they were gone, went back to the beach and bought more. Some did not bother to shoot them, stripping the saddles off and the shoes and leaving them where they fell. Their hearts turned to stone—those which did not break—and they became beasts, the men on the Dead Horse Trail."

On that trail in the winter of 1897-1898 no fewer than 3,000 horses perished. Even as the animals collapsed, a better way up the White began building. The first rails of the narrow-gauge White Pass Railway were laid at Skagway in May of '98 and, with 35,000 men ripping away cliff and hillside with picks and blasting powder, reached the summit 21 miles away in a little over eight months. By July of 1899, the rails had reached all the way to Lake Bennett, but by then the great gold rush was as dead as the horses.

Men too poor to buy horses took the Chilkoot Trail from Dyea. The Chilkoot was shorter than the White Pass Trail, but the distance was deceptive. Of the two routes, it was the more harrowing by far. In the final four miles, the grade was 30 per cent and in the last half mile 35 per cent, so that men had to climb leaning forward, in lockstep.

In winter sled dogs could be used to traverse much of the trail, carrying 10 to 15 pounds apiece in little saddlebags, although even they had to be carried up the last terrible slope. These were real sled dogs—Huskies and Malemutes, bred to the task and highly prized. A good one was worth $350, and they were well fed and cared for. When a sled dog began to limp, with ice or snow packed between the pads of its feet, its owner would take the dog's foot in his mouth to melt the ice, then carefully wipe the foot dry on his shirt.

There were also human packers on the trail, Chilkat Indians who often worked with their families. A rugged man could carry as much as 150 pounds, his wife 75 and his children anywhere from 26 to 60. When the first Klondikers began to use the trail the In-

A carpet of dead and dying horses covers the Skagway Trail to the Klondike in 1897. Over 3,000 starved, overburdened animals died en route; men swore some of them committed suicide leaping from cliffs.

dians charged five cents a pound, but after the stampede began the rate quickly went to 40 cents.

The Chilkoot, like the White Pass Trail, had its swamps and precipices, but it was that final, four-mile ascent that all men remembered as long as they lived. It was there that the weak of spirit or body quit and went home, and there that some of the most extraordinary feats of the gold rush were performed. For much of the year the ascent was covered with snow and ice, but even in midsummer the climbing was terrible because the trail became a mass of mud and slippery shale. A half mile below the summit there was a great flat ledge called The Scales, where all baggage was reweighed and the rates of the packers went up to a dollar a pound. Although the Klondikers had long since discarded their nonessentials, each still had close to a ton of necessities to be taken over the cruel pass.

At The Scales the truth hit them: the average load that a city-bred man or woman could carry up that slope was 50 pounds, and it took six hours to make the climb. At that rate it would require 35 or 40 trips and—allowing for delays caused by frequent storms—three months to get the job done. Nor could anyone cheat on the amount of food. At the summit on the international border there was a North-West Mounted Police post. The Mounties, who were polite but firm, would not permit anyone to enter the Yukon wilderness with less than a half year's rations.

Early in the rush the ascent was without any improvements, just as nature had constructed it. Then two partners with axes hacked out steps in the ice near the top and charged a toll for the use of them; even-

A team of Angora goats is pressed into service hauling supplies to the Klondike. So desperate was the demand for sled teams and pack animals that virtually anything with four legs might find itself on the trail: sheep, burros, dehorned reindeer, and mongrel dogs of every size and description.

tually there were 1,500 steps in the mountainside. By December 1897 there were rope cables and in 1898 a gigantic steam-driven tram with 30 miles of steel cable on a monster drum. But before the tram, according to records kept by the Mounties, 22,000 people got over the Chilkoot on foot. Each carried his burden to the summit, put it down in a dump that covered many acres and marked it with a pole from which fluttered a rag. He then tucked a canvas pad under him and careened downhill in a grooved, shoulder-deep trail like a bobsled run. Only the rarest, strongest men could make more than one round trip a day.

But there were several rare, strong men. One man lugged, in sections, a piano to the summit. A man from Iowa, A. J. Goddard, and his wife somehow managed to carry over the disassembled parts of two small stern-wheel steamboats and put them back together in the Yukon headwaters. There were people who, considering their burdens, seemed insane—a man struggling with a heavy grindstone, another carrying crates of cats. There was a contest between an Indian packer called Jumbo and a big Swede named Anderson. On a bet each made his way to the summit loaded with a staggering 300 pounds. When they returned to The Scales the Swede, still strong, picked up a second 300-pound burden and gave the Indian a challenging glance. The Indian shook his head and turned away.

Most of the Klondikers in 1897 got across the passes too late in the fall to go on to Dawson; the lakes and rivers were frozen. Therefore they settled down in encampments along the way. The largest of these was a tent city on the shore of Lake Bennett, where some 10,000 people wintered amid great piles of supplies. The average miner, so far from home, might have given way to despair. He did not. He had crossed his high pass in the mountains, carrying his ton of goods, and he was not the same man he had been when he left home. He did make an occasional outcry, but less in sorrow than because of the work he had to do. He had to build a Godforsaken boat in order to cross the lake and navigate down the Yukon to the Klondike.

All winter long, while they waited for the ice on the lakes and the rivers to break up, the Klondikers sawed their green, irregular, warping planks from the surrounding spruce forests, and set about building one of history's most incredible flotillas. Most of them knew nothing about naval architecture, but they exchanged information and they did their best. The sound of hammering was incessant during the short winter days.

The boatbuilders at Lake Bennett had little or no knowledge of what was going on at Dawson City. In only a year it had become a boomtown populated by 6,000 people who had been smart enough and fast enough to get there ahead of the grand stampede. Some had come up from the mouth of the Yukon River on the little steamers; others had crossed the White or the Chilkoot very early; still others, in response to the mysterious summonings of gold, had simply walked in out of the moose pasture or come drifting down the network of rivers and creeks.

By the summer of 1897, when most of the Klondikers were still buying their outfits in Seattle or Portland, Dawson already had 300 buildings and 10 saloons, none of them taking in less than $2,000 a week. Payment for drinks was naturally made in gold, which was handled so casually that a teen-aged boy made $278 over the course of a few weeks simply by panning the sawdust on the floor of a single bar.

At that point, a good many men were still searching for likely places to stake claims, and finding fewer and fewer of them. In the already-established mines, gold was coming to the surface in buckets as men crawled along the drifts, scraping up dirt so rich that a single shovelful could yield $800. But almost as fast as it came up, the gold was exchanged for goods and services at astronomical prices. In the dance halls the miners happily paid one dollar for a whirl around the floor that lasted about a minute. Their partners were the same sturdy, homely types found in all Western mining towns from 1850 onward, differing only in name in the Klondike: Dog-faced Kitty, the Grizzly Bear, Nellie the Pig, the Oregon Mare and Diamond-Tooth Gertie. The girls—merely for dancing and pushing drinks—made $100 a night. The prostitutes earned more, but they, too, were pinched by the general inflation. Kerosene for the red lanterns outside their doors cost $40 a gallon.

There were acute shortages of other items, often the most ordinary ones. It cost $150 a day to rent a horse and wagon, and hay had gone to $1,000 a ton. Salt sold for its weight in gold. The only broom for sale in town fetched $15. But the greatest lack was food.

Most of the early comers to Dawson had been living in the region or had arrived before the North-West Mounted Police began to enforce the rule of a half year's rations. Now, as the winter of 1897-1898 closed in, the Mounties urged people without ample food stocks to leave town, perhaps by going downriver to Circle City or Fort Yukon. A few hundred men did so, but the rest remained. No one starved, although there were numerous cases of scurvy among miners living on unvaried diets of flapjacks and beans.

As winter softened into spring, the boatbuilders at Lake Bennett, and at other camps strung out for 60 miles along the headwaters of the Yukon, waited impatiently for the ice to break up. The first cracks appeared on Bennett on May 29, 1898, and within two days the waterway was open and the boat race was on.

Although an odder fleet may have put out from shore somewhere, sometime, it is hard to find record of it. All told, some 20,000 people embarked in more than 7,000 boats of every conceivable design—square boats, round boats and triangular boats, catboats and flatboats, dories and wherries, rowboats and canoes, kayaks, outriggers, one-man log rafts and big scows that could carry 10 men and their gear. There was a great tonnage of cargo, including the equipment and cash required to open branches of two leading Canadian banks and the printing presses of two newspapers. An Italian fruit merchant embarked with eight tons of oranges, lemons, cucumbers and other produce that he had laboriously brought in over the mountains.

In a fresh breeze the Klondikers hoisted their sails, some of canvas and some of blankets or coats, and in the case of two ladies traveling together, of billowing bloomers and slips. Down Lake Bennett they all went in *Gussie, Pearl* and *Lizzie, Lucky Star, Four Leaf Clover* and *Golden Horseshoe*. At dusk on the first day the breeze ceased and all the boats were becalmed. Somewhere in the silent fleet a voice began to sing, and soon hundreds of songs were floating across the wide, dark lake.

Below Bennett there were two more mountain lakes connected by short channels; then the waterway, narrowing, entered a vicious stretch containing Miles Canyon, Squaw Rapids and White Horse Rapids. In the past, experienced men had lost their lives here among the rocks and whirlpools, and now the Klondikers, innocent and generally unprepared, were swept into the maelstrom in their leaky craft. In the first few days 150 boats were wrecked and 10 men drowned. Above the rapids a colossal traffic jam developed; thousands of boats were beached while their owners tried to decide what to do. At this point the North-West Mounted Police, who had been watching with concern, laid down some safety laws; women must walk around the rapids (five miles), and no boat could go through until the Mounties were satisfied that it had been made rapids-worthy, its cargo had been properly stowed, and a competent man was steering. Thereafter only a few wrecks and drownings occurred, although the passage remained fearsome for all who made it.

One of the professional pilots taking gold-rushers' craft through the fast water at fees ranging up to $100 was Jack London. The future novelist worked at the rapids long enough to accumulate $3,000 in pilotage, which was considerably more than he ever made digging gold out of the ground.

Not surprisingly, considering the variety and clumsiness of the craft and the wildly varying degrees of seamanship among the owners, the fleet was soon strung out all along the 500 miles from Bennett to Dawson. It was spring, and lupines, bluebells and briar roses bloomed on the banks. The midnight sun shone 22 hours of every day. And it shone on scenes of adventure and misadventure for nearly every mile.

A Norwegian in a big bateau blithely essayed to shoot the rapids without a paddle while playing his music box. He kept it up until the waters, compressed between the canyon walls, rose up in a great surging lump and swamped him.

Even on placid Lake Bennett things could go amiss. An 18-year-old bride, Mrs. Mabel Long from California, was aboard a scow with her 38-year-old husband and a young Bostonian named Rossburg, when the craft ran up on a rock and she was pitched overboard. While Long jittered, wrung his hands and yelled for help, Rossburg jumped overboard and hauled the drowning bride to safety. As soon as she had recovered her composure and dried out, Mabel announced that she was leaving her husband. She had detested him anyhow and had married him only because her parents insisted. And so, pursued by an angry spouse, she

ran away with Rossburg and was later pleased to learn that her paramour was heir to a Boston fortune.

Friendships and solemn partnerships dissolved in anger. Two men marooned on rocks in the middle of Thirty Mile River were seen trying to kill each other with their bare fists. Ten men beached their big scow at Big Salmon, divided all their gear into separate piles on separate blankets, then solemnly dismembered the scow and built 10 little scows out of its planks. A Mountie spent a whole day futilely trying to reconcile six preachers bent upon dissolving their association because, as they complained, they had fallen into the sin of telling unchristian lies to one another.

The run from Lake Bennett to Dawson took about two weeks. By mid-June the full force of the rush hit the mining town and boats were tied up, in many places two or three deep, for nearly two miles along the bank of the Yukon. Scrambling ashore, the Klondikers swarmed along Dawson's Front Street. They crowded into the saloons, asking oldtimers, newcomers, anyone, the direction of the gold creeks and how long it might take a man to make, say $20,000.

The town where they had arrived, after 10 months of brutal strife, was like nothing any of them had ever seen before—and it was changing so fast that it became unrecognizable from week to week. While the small boats had been running down the Yukon from Lake Bennett, steamers were coming up from the Bering Sea carrying more gold seekers and more thousands of tons of everything. During a few weeks in June 1898, Dawson's population grew to about 28,000, very close to that of Portland or Seattle. The choicest corner lots on Front Street, suitable for saloons, went for $40,000 and two sawmills worked 24 hours a day turning out building materials. There were four churches, a brewery, and halls for the Masons and the Odd Fellows. By the end of the year, Dawson had telephones, electric lights and moving pictures.

Big fires struck the town in 1897, 1898 and again in 1899. But each time it was swiftly, more solidly and expensively rebuilt. Streets of tents gave way to cabins and cabins yielded to houses. Before long houses had parlors complete with pianos and carpets fetched upstream from St. Michaels on the steamers.

The same steamers brought cargoes to stock the thriving stores along Front Street with almost anything

Rigged up as Yankee Doodle, with a burro serving in lieu of a pony, a north-country patriot gets set outside the flag-bedecked Skagway City Hall for the 1898 Fourth of July parade. Across the border in Dawson, the American holiday was celebrated along with Canada's Dominion Day.

the heart could desire: cigars, hot-house grapes, ostrich feathers and opera glasses, plug hats and boiled shirts, patent leather shoes, paperback novels and jewelry.

Every night became an all-night revel. In theater saloons like the Monte Carlo, miners listened to plays such as *Camille* until the curtain came down and then, to the tunes of an organ, spent the rest of the night cavorting with the girls on the dance floor. Meanwhile, upstairs in the balcony of private boxes, the really affluent swells bought champagne for their girl friends at $40 the quart; one big spender set something of a record by ordering $1,700 worth of champagne for a bevy of beauties in a single session. A customer who wanted to gamble was accommodated, first depositing his poke with the dealer; when he was either broke or satisfied he got the poke back minus his losses; if he won he got it back full, along with chips representing his profits.

A roll call of the citizenry abounded in names that would one day become famous: Alexander Pantages, the future motion picture magnate; Tex Rickard, already a fight promoter; Sid Grauman, whose celebrated Chinese Theater was yet to come; Augustus Mack, who would build the trucks that bore his name; Key Pittman, who would one day become a powerful senator from Nevada.

And the resident Midases, whose reputations were already made, were spectacularly visible. They were men such as Fred Bruceth, who had panned $61,000 in a single day; Lucky Swede Anderson, who was conned into paying $800 for a supposedly worthless claim, and then dug one million dollars in gold from it; Big Alex McDonald, whose claim at Eldorado *Thirty Six* had yielded the largest Klondike nugget, a chunk of gold that weighed three and a half pounds and was worth almost $600.

Swiftwater Bill Gates, a wispy onetime Circle City dishwasher, had made his first pile by joining a syndicate of seven men who took a chance on Eldorado *Thirteen,* braving the jinx inherent in the number. Seven test holes later they were all rich men and Bill, now in plug hat, Prince Albert and diamond stickpin, was off and away. Bill never drank — although he sometimes toned up his five-foot-five body with a dip in a bathtub full of wine — but he became a nightly, all-night box habitué at the Monte Carlo. When he gambled he would throw his poke on the faro table and announce: "Raise 'er up as far as you want, boys, and if the roof's in your way tear it off!" He developed a consuming appetite for girls, the younger the better. The first one was Gussie Lamore, a 19-year-old temptress who had an insatiable appetite for eggs. It was because of her that Bill picked up another admiring nickname, the Golden Omelet, by cornering Dawson's entire stock of this nearly priceless commodity.

Later, on a junket outside to Seattle, he enticed away the convent school daughters, Bera, 15, and Blanche, 19, of Mrs. Iola Beebe, who had been trying to entice *him* into setting her up in business. In a runaway elopement from Skagway, he wed the 15-year-old, got her pregnant and not long after, with a roving eye ever on the cradle, ran away with and married a 17-year-old, Kitty Brandon, without bothering to divorce Bera. Meanwhile, with the devil's own luck, he kept striking it rich and blowing it even more richly. Years later he died at 67 in Peru, the operator of what was reported to be a vast silver-mining concession.

Of all the characters in Dawson, none attracted more fascinated attention than Big Alex McDonald. Huge, ham fisted, so taciturn as to seem nearly beyond the reach of human communication, Alex perfected, if he did not actually invent, the "lay" system. This meant that he did not himself dig for gold, but bought claims and then leased them out to working miners for a percentage of the results. He bought, traded, leased, mortgaged and, to finance his transactions, sometimes borrowed money at interest rates as high as 365 per cent per annum. He was often stupefyingly in debt, but that did not bother him. Once when he was down $150,000 in short-term loans, he said, "I can dig out $150,000 any time I need it." And he could; at one time he owned pieces of 65 of the richest creek claims in the Klondike and many more on the benches.

Not even Big Alex knew how wealthy he might be at any moment; once he spent an entire afternoon trying to recall his assets for the Canadian Bank of Commerce, from which he needed some working capital. To Big Alex the gold itself was "trash"; his peculiar passion was to control workings of the mines — the land under which gold might lie and the men who dug it. On the sideboard at home he kept 45 pounds of nuggets in a bowl. "Help yourself. Take some of the bigger ones," he told a newspaperwoman who interviewed him. She hesitated but Alex waved her on. "Take as many as you please. There are lots more."

This was the changeling town on which the horde of Klondikers made their landfall in the summer of 1898, and men like Swiftwater Bill and Big Alex were its dazzling exemplars. For a little while, on that frenetic waterfront, it was still possible to believe in the dream that, as recorded by a 17-year-old rusher from San Diego, "all they would have to do was to pick the nuggets above the ground and some even thought they grew on bushes."

But then, slowly and with varying degrees of reluctance, Klondikers accepted the truth: there was no good ground left to claim and there were thousands to claim it. Some swore, some got drunk, most did both. A correspondent for *Harper's Weekly* looked at them with a compassionate eye: "Who is there that can de-scribe the crowd, curious, listless, dazed, dragging its way with slow, lagging step along the main street? Can this be the 'rush' that newspapers are accustomed to describe as the movement of gold seekers? Have the hard, weary months of work on the trails exhausted their vitality? Or is it the heavy shoes that make them drag their feet so wearily along the street?"

A few were still trying, but hope was ebbing. "Our time in Dawson was spent out in the country, travelling up and down the creeks hunting for a location, but every likely spot within 50 miles of Dawson is staked out," wrote a rusher. "There is always a crowd of men waiting outside the recorder's office. I waited from Monday to 3 p.m. Friday before my turn came. Finding that there was apparently not the slightest chance to get anything I decided to return home."

For some, of course, Dawson itself was the gold mine. The Italian fruit merchant sold his oranges and

Solitaire gives comfort to a lonely miner. Klondikers' favorite solace was whiskey: 120,000 gallons were brought to Dawson in 1898.

lemons for one dollar apiece, and tomatoes for five dollars a pound. A competitor brought in a watermelon and got $25 for it. Some of the men who had appeared mad back at White Pass or on the Chilkoot now seemed eminently sane. The man with the grindstone, who knew what hacking at frozen ground can do to metal, soon had customers waiting in line to have their picks sharpened at an ounce of gold apiece. The man with the cats, who understood the loneliness of miners in isolated cabins, got an ounce of gold apiece for them, too.

Some Klondikers were able to get jobs in the sawmills and warehouses or worked as woodcutters, porters or watchmen. Others went out to the creeks and signed on as pick and shovel laborers in the mines.

Though they could not possess it, they could at least see and touch the gold as the piles of rich dirt, built up during the winter, were washed in wooden sluices.

But for most of the Klondikers there was little to do but mill around in the crowded, muddy streets. They had only what was left of their flour, beans and bacon, their battered little boats, and their picks, shovels and gold pans. In the aimless throng former partners, who had quarreled on the trail and split up, encountered each other and shook hands with sheepish grins. But it was seldom a simple matter to locate an old friend whose good points you now suddenly remembered. "The main street is always crowded with men trying to find one another . . . for it is a hard matter to find a man in Dawson and much time is wasted thereby," Ar-

Candles in hand, a bevy of Klondikers take their ease on the hillocks of frozen gold-bearing gravel in a long-buried stream be

thur Christian Newton Treadgold wrote to the *Manchester Guardian*. "When you find your man the two of you sit on the edge of the sidewalk (raised a foot above the road for cleanliness) and talk." Now and then the effort at reconciliation came too late. One man spent weeks looking for a missing partner without success. Then, asked to serve as a pallbearer at the funeral of an unknown typhoid victim, he looked into the coffin and recognized his man.

In these enforced doldrums, caught in an idleness they could never have indulged in all those frantic months on the trail, men spent a great deal of time in gossip, in the saloons, in tents, aboard their moored boats. The Spanish-American War was in full swing, and when recent copies of the Seattle *Post Intelli-*

gencer reached town there was sharp competition for them. One issue, containing an account of Admiral Dewey's victory in Manila Bay, was bought at auction by a miner for $50. He then hired a loud-voiced lawyer to read it in a public hall, charged one dollar admission and made about $1,000.

After a few weeks in Dawson, sometimes after only a few days, many Klondikers started for home. In August 1898, Front Street was piled with miners' outfits being offered for sale. So much stuff was sold so fast that the price of flour dropped to less than the owners had paid for it in Seattle. Having no choice, the Klondikers took whatever they could get, crowded aboard one of the river steamers and sailed off down the Yukon. The river was nearly as jam packed with de-

The tykes in the foreground were not there on a visit: at the peak of the rush, every able-bodied man, woman and child was busy digging.

parting miners as the Chilkoot had been coming up. One downriver boat, the *W. K. Merwyn,* was so overloaded, a passenger reported, that they "had to stand like straphangers in a street car." The *Merwyn* ran out of food and had to tie up frequently while the ship's company hunted for goose eggs on the shore.

And so the great adventure ended. Turning homeward, some recognized at last that if there had not been many winners in the Klondike gold rush, the experience had not produced many real losers either. As the correspondent from *Harper's* wrote at the end: "The final adieus to friends and companions in hunger and plenty, in misery and good fortune, were a fitting close to sixteen months of an experience that none of us can hope to see repeated in a lifetime. A life of free-

dom and adventure has a fascination that grows rather than diminishes, and yet the privations that every person who went into the Klondike endured taught him better to separate the good from the bad, the essential from the non-essential and to recognize the real blessings and comfort of civilization."

The statistics of the Klondike rush can never be precise, but a fair estimate is that of the 100,000 people who set out for Dawson, 30,000 to 40,000 actually got there. Only half of them bothered to look for gold. Perhaps 4,000, mostly oldtimers, found some, but only a few hundred got enough to be considered rich—and even of these, all but a handful soon gambled, drank or otherwise tossed away their money. Dawson was a metropolis for a single year, from midsummer 1898 to

Shovelers clear the blizzard-bound tracks of the White Pass and Yukon Railroad, built between Skagway and Lake Bennett in 1899.

midsummer 1899. It continued to be an important mining center for several years thereafter, when large corporations and big dredges moved in to scoop up most of the gold—a grand total of about $300 million. But after the turn of the century the excitement was gone and the population dwindled.

Ambrose Bierce, who in the late 1890s was a sour-minded journalist on the San Francisco *Examiner,* had a low opinion of the stampede to the Klondike. "But the blue-nosed mosquito-slapper of greater Dawson, what is he for? . . . Nothing will come of him. He is a word in the wind, a brother to the fog. At the scene of his activity no memory of him will remain. The gravel that he thawed and sifted will freeze again. In the shanty that he builded, the she-wolf will rear her poddy lit-ter, and from its eaves the moose will crop the esculent icicle unafraid. The snows will close over his trail and all be as before."

No doubt Bierce was right, at least in the strict phys-ical sense. All gold seekers vanish and all their cabins turn to dust. But there is something more to be said about those who took part in the last great rush. Among the people who went over the White Pass or climbed the Chilkoot, sawed their own planks, built their own boats, ran the rapids and finally set foot in Dawson, the disappointment in not finding gold was eventually forgotten. They found something within themselves that they prized more highly, and it warmed and sustained them until they were old, worn men and women and went down valiantly into the earth.

At last a locomotive gets through, running in a cut as deep as its headlight. The tracks climbed 2,885 feet to the summit in just 21 miles.

Roisterous times in Baghdad on the Yukon

To the weary Klondikers who finally reached Dawson City in the spring and summer of 1898, the place seemed nothing short of Baghdad on the Yukon. Imagine a town where the main bank kept a million dollars in gold dust in two tin-lined wooden boxes without a lock and without a guard!

The gold belonged to only a fortunate few, of course. Yet, even for the most broke and most disappointed gold rusher there was excitement—though it was often only the excitement of seeing how more fortunate people spent their money. There was, for example, Sailor Bill Partridge, who changed clothes several times a day and never wore the same suit twice. Or Charlie Kimball, who went on a drunk that lasted three months and cost $300,000. Or Silent Sam Bonnifield who risked $75,000 in one pot of stud poker. There was also Coatless Curly Munro's championship dog team, groomed on $4,320 worth of bacon, fish and flour.

For many, the adventure itself was a priceless treasure. Will Langille was as unsuccessful at finding gold as most of his fellows. But he summed it up, not only for those who raced headlong to Dawson but for all gold rushers, when he exulted: "It was there, and I would not have missed it for anything."

A river steamer lies at the quay in the broad Yukon at Dawson in 1898. A smaller stream, the Klondike, bisects the town from the east.

226

Sunk to the hubs in mud, a load of lumber bogs down on Dawson's Front Street. An unusually hot spring in 1898, with temperatures as high as 110 degrees, brought rising waters and an early breakup of the Yukon River's ice, flooding the bottom lands on which the town was being built.

A mob jams a Dawson street in the summer of 1898 to hear a strong-voiced townsman *(left center)* read a rare state-side journal. Miners were particularly avid for news of the Spanish-American War, and word of the July 3 defeat of a Spanish fleet in Cuba was greeted with thunderous cheers.

Gutted by fire on April 26, 1899, Dawson's Opera House stands in charred ruins the morning after. The blaze, which broke out in a dance-hall girl's bedroom, destroyed much of the town's central business district. Men had to fight it with dynamite after hoses froze and burst in the –45° cold.

231

Taking on a standing-room-only deckload of passengers, a Yukon River steamer prepares to leave Dawson. By 1899, the Klondike rush was over, and 8,000 people left town in the month of August alone — many of them heading home, but some hurrying off to a new gold find in Nome.

TEXT CREDITS

For full reference on specific page credits see bibliography.

Chapter 1: Particularly useful sources for information and quotes in this chapter: Vardis Fisher and Opal Laurel Holmes, *Gold Rushes and Mining Camps of the Early American West,* The Caxton Printers, Ltd., 1968; LeRoy R. Hafen, ed., *Pike's Peak Gold Rush Guidebooks of 1859,* The Arthur H. Clark Company, 1941; Watson Parker, *Gold In the Black Hills,* University of Oklahoma Press, 1966; Rodman Wilson Paul, *Mining Frontiers of the Far West, 1848-1880,* University of New Mexico Press, 1974; Glenn Chesney Quiett, *Pay Dirt,* D. Appleton-Century Company, 1936; T. H. Watkins, *Gold and Silver in the West,* American West Publishing Company, 1971; Otis E. Young Jr., *Western Mining,* University of Oklahoma Press, 1970; 22—William Parsons quotes, Hafen, p. 155; 35, 37—Kellogg quote, *Jim Wardner,* p. 58; 37—Rickard quote, Rickard, *A History of American Mining,* p. 322; 40—Wickersham quotes on miners' justice, O'Connor, pp. 162-163; Custer City description, Smith, *Rocky Mountain Mining Camps,* p. 42. Chapter 2: Particularly useful sources for information and quotes: Dan De Quille, *The Big Bonanza,* Alfred A. Knopf, 1947; William S. Greever, *The Bonanza West,* University of Oklahoma Press, 1963; Eliot Lord, *Comstock Mining and Miners,* Howell-North, 1959; Grant H. Smith, *The History of the Comstock Lode, 1850-1920,* Nevada Bureau of Mines & University of Nevada, 1943. Chapter 3: Particularly useful sources for information and quotes: J. W. Dilley, *History of the Scofield Mine Disaster,* publisher unknown, circa 1900; Richard E. Lingenfelter, *The Hardrock Miners,* University of California Press, 1974; Clark C. Spence, *Mining Engineers & the American West,* Yale University Press, 1970; Otis E. Young Jr., *Black Powder and Hand Steel,* University of Oklahoma Press, 1970. Chapter 4: Particularly useful sources for information and quotes: Otis E. Young Jr., *Western Mining,* University of Oklahoma Press, 1970; Glenn Chesney Quiett, *Pay Dirt,* D. Appleton-Century Company, 1936; T. A. Rickard, "Salting," *Engineering and Mining Journal,* March 1941; 137—Mark Twain quote on assayer, *Roughing It,* p. 195; 137-138—Mark Twain on publicity, *Roughing It,* pp. 231-232; 142—Mark Twain quote, Quiett, pp. 20-21; 143—Hittell quote, Watkins, pp. 248,258; 143-144—bilking stockholders quote, De Quille, pp. 345-346; 144—ensuring secrecy, Lord, pp. 288-289; De Quille on secret shifts, *The Big Bonanza,* pp. 309-310; 148—Sunday service quote, Crampton, p. 50; Rickard quote, *Mining and Scientific Press,* June 6, 1908, p. 776; 148-151—Richardson on Denver House, Fisher and Holmes, pp. 193-194; 151-153 —Canfield quote, Chafetz, pp. 119-120; 153—miners and women quote, Leighton, p. 106; Molly b'Damn description, *Life,* May 11, 1959, p. 66; poem about Mollie May, Smith, *Rocky Mountain Mining Camps,* p. 230; 157—Mattie Silks description, Fisher, p. 208. Chapter 5: Particularly useful sources for information and quotes: Edward Blair, *Palace of Ice,* Timberline Books, 1972; John Burke, *The Legend of Baby Doe,* G. P. Putnam's Sons, 1974; Oscar Lewis, *Silver Kings,* Alfred A. Knopf, 1967; Glenn Chesney Quiett, *Pay Dirt,* D. Appleton-Century Company, 1936; Duane A. Smith, *Horace Tabor: His Life and the Legend,* Colorado Associated University Press, 1973; George F. Willison, *Here They Dug the Gold,* Reynal & Hitchcock, 1946; 164 —Geological survey quote, Lord, p. 311; 174—Ice Palace description, Harvey, p. 97; 175—Description of Baby Doe, Fisher and Holmes, p. 381. Chapter 6: Particularly useful sources for information and quotes: Pierre Berton, *The Klondike Fever,* Alfred A. Knopf, 1958; Vardis Fisher and Opal Laurel Holmes, *Gold Rushes and Mining Camps of the Early American West,* The Caxton Printers, Ltd., 1968; William S. Greever, *The Bonanza West,* University of Oklahoma Press, 1963; Richard Mathews, *The Yukon,* Holt, Rinehart & Winston, 1968; Richard O'Connor, *High Jinks on the Klondike,* Bobbs-Merrill, 1954; Kathryn Winslow, *Big Pan-Out,* W. W. Norton, 1951; 210-211—Jack London quote, Walker, pp. 69-70; 222—*Harper's* correspondent quote, Adney, pp. 455-456.

PICTURE CREDITS

The sources for the illustrations in this book are shown below. Credits from left to right are separated by semicolons, from top to bottom by dashes.

Cover—Courtesy The Bettmann Archive. 2—Courtesy The Bancroft Library. 6,7—Courtesy Museum of New Mexico. 8,9—Courtesy Idaho Historical Society. 10,11—Courtesy Western History Dept., Denver Public Library. 12,13—Courtesy Nevada Historical Society. 14,15 —Courtesy Arizona Historical Society. 16,17—E. A. Hegg, courtesy Photography Collection, Suzzallo Library, University of Washington. 18 —Courtesy Library, State Historical Society of Colorado. 20,21—Courtesy Western History Dept., Denver Public Library. 23—Map by Robert Ritter. 24—Courtesy Library of Congress. 25—Courtesy Utah State Historical Society. 26—Lloyd Rule, courtesy Library, State Historical Society of Colorado. 28,30,31—Courtesy Western History Dept., Denver Public Library. 34,35—Courtesy Library of Congress. 36 —Courtesy Library, State Historical Society of Colorado. 38,39—Courtesy Montana Historical Society, Helena. 41—Courtesy Library, State Historical Society of Colorado. 43—Courtesy Montana Historical Society. 44,45—Courtesy National Archives, #77-HQ-264-854. 46,47 —Stanley J. Morrow, courtesy J. Leonard Jennewein Collection, Layne Library, Dakota Wesleyan University, Mitchell, S.D. 48,49—Courtesy Deadwood Public Library, Deadwood, S.D. 50,51—Courtesy National Archives, #111-SC-100863. 52,53—Courtesy Nebraska State Historical Society. 54,55—Reproduced by permission of The Fine Arts Museums of San Francisco. 56—J.H.Crockwell, courtesy Nevada Historical Society. 58,59—Courtesy Library of Congress. 60—Courtesy Nevada Historical Society. 61—Courtesy Special Collections Dept., University of Nevada, Reno, Library. 62,63—Courtesy The Bancroft Library. 64, 65—Courtesy Mackay Museum, Mackay School of Mines, University of Nevada, Reno. 66,67—Courtesy The Bancroft Library. 68—Timothy O'Sullivan, courtesy National Archives, #77-KS-1-13. 69—Timothy O'Sullivan, courtesy National Archives, #77-KW-140. 70,71—Timothy O'Sullivan, courtesy National Archives, #77-KS-1-17—Timothy O'Sullivan, courtesy National Archives, #77-KW-3-139; Timothy O'Sullivan, courtesy The Bancroft Library. 73—Courtesy Special Collections Dept., University of Nevada, Reno, Library. 74,75—Background, C. E. Watkins, courtesy The Bancroft Library, insets, C. E. Watkins, courtesy California State Library (2)—courtesy The Bancroft Library; courtesy Nevada Historical Society. 77—Courtesy Nevada Historical Society. 78,79—Courtesy California Historical Society, San Francisco/San Marino. 80 through 91—C. E. Watkins, courtesy California State Library. 92—Courtesy The Bancroft Library. 95—Courtesy Amon Carter Museum, Fort Worth, Texas. 96,97—W. H. Jackson, courtesy

Richard A. Ronzio Collection. 98—Courtesy Western History Dept., Denver Public Library. 100,101,102—Courtesy Library, State Historical Society of Colorado. 103—Courtesy Western History Dept., Denver Public Library. 105—Arnold F. Olean, Gold Mine Silver Screen, Black Hawk, Colorado—courtesy Western History Dept., Denver Public Library. 106,107—Courtesy Western History Dept., Denver Public Library. 108—Dean Austin, courtesy Mackay Museum, Mackay School of Mines, University of Nevada, Reno. 110,111—Courtesy The New York Public Library, Astor, Lenox and Tilden Foundations. 112—Courtesy Special Collections, University of Arizona Library. 113—Courtesy Homestake Mining Company, Public Affairs Dept., Lead, S.D. 114 —Courtesy Smithers Collection. 116,117—Courtesy Amon Carter Museum, Fort Worth, Texas. 118—Courtesy Western History Dept., Denver Public Library. 119—Courtesy Idaho Historical Society. 121 —Courtesy TIME-LIFE Picture Agency. 122 through 133—George Edward Anderson, from the collection of Robert W. Edwards. 134,135 —Courtesy Western History Dept., Denver Public Library. 136 —Courtesy Nevada Historical Society. 137—Courtesy Seattle Historical Society. 138 through 141—Courtesy Special Collections Dept., University of Nevada, Reno, Library. 143—Courtesy Western History Dept., Denver Public Library. 145—Stanley J. Morrow, courtesy the W. H. Over Museum, Vermillion, S.D. 147—E. A. Hegg, courtesy Photography Collection, Suzzallo Library, University of Washington. 149—Courtesy Amon Carter Museum, Fort Worth, Texas. 150,151 —Courtesy Nevada Historical Society. 152—Courtesy the W. H. Over Museum, Vermillion, S.D. 154,155—E. A. Hegg, courtesy Photography Collection, Suzzallo Library, University of Washington. 156

—Courtesy Special Collections Dept., University of Nevada, Reno, Library; courtesy Nevada Historical Society. 158,159—Courtesy California Historical Society, San Francisco/San Marino. 160—Courtesy Nevada Historical Society, except bottom, courtesy Mackay Museum, Mackay School of Mines, University of Nevada, Reno. 163—Courtesy Library of Congress. 165—Courtesy Nevada Historical Society. 166, 167—Courtesy Special Collections Dept., University of Nevada, Reno, Library. 170,171—Dean Austin, courtesy Special Collections Dept., University of Nevada, Reno, Library. 173—Courtesy The Bancroft Library. 176 through 181—Courtesy Library, State Historical Society of Colorado. 182,183—Courtesy Western History Dept., Denver Public Library. 184 through 187—Courtesy Library, State Historical Society of Colorado. 188,189,192-197,212,213,216,217,220-223,226, 227,230,231—E.A.Hegg, courtesy Photography Collection, Suzzallo Library, University of Washington. 190,191—Asahel Curtis, courtesy Seattle Historical Society. 198—Ron Klein, courtesy Robert DeArmond, Alaska Historical Library, Juneau. 200,201—Courtesy Seattle Historical Society. 203—Map by Robert Ritter. 204,205—F. LaRoche, courtesy Photography Collection, Suzzallo Library, University of Washington. 208,209—E.A.Hegg, courtesy Photography Collection, Suzzallo Library, University of Washington, except top left, F. LaRoche, courtesy Photography Collection, Suzzallo Library, University of Washington. 211—Asahel Curtis, courtesy Seattle Historical Society. 219 —Courtesy Photography Collection, Suzzallo Library, University of Washington. 224,225,228,229—Asahel Curtis, courtesy Seattle Historical Society. 232,233—Asahel Curtis, courtesy Photography Collection, Suzzallo Library, University of Washington.

ACKNOWLEDGMENTS

The index for this book was prepared by Gale L. Partoyan. The editors wish to give special thanks to Dr. William S. Greever, Professor of History, University of Idaho, who read and commented on portions of the text. The editors also thank the following: Prof. John F. Abel Jr., Prof. Charles Frush, Department of Mining and Engineering, Colorado School of Mines, Golden; Dr. Maxine Benson, Virginia Roberts, Ellen Wagner, Colorado State Historical Society, Denver; Richard Berner, Head of Archives, Ellen Guerricagoitia, Sandra Kroupa, Robert Monroe, Carol Zabilski, Special Collections, University of Washington, Seattle; Edward Blair, Leadville, Colo.; Fred C. Bond, Consultant, Allis-Chalmers Corp., Tucson, Ariz.; Margaret Bret Harte, Ref. Librarian, Loretta Davisson, Asst. Ref. Librarian, Arizona Historical Society, Tucson; Kenneth J. Carpenter, Librarian, Carrie Townley, Special Collections, University of Nevada, Reno; James H. Davis, Photo Archivist, Idaho State Historical Society, Boise; Robert De Armond, Juneau, Alaska; Zelma Doig, Librarian, Alaska State Library, Juneau; Robert W. Edwards, Salt Lake City, Utah; Suzanne Gallup, Ref. Librarian, Irene Moran, The Bancroft Library, University of California; Eleanor Gehres, Director, Western History Collection, Denver Public Library, Denver; Augustino Mastrogiuseppe, Picture Librarian, Elaine Gilleran, Director, Wells Fargo Bank History Room, San Francisco, Calif.; Mrs. Helene Glocer, Public Relations Asst., Anaconda Company, New York; Alec Hansen, Director of Public Affairs, Anaconda Company, Butte, Montana; Mrs. Virginia Herald, Arthur Lakes Library, Colorado School of Mines, Golden; L. James Higgins, Nevada State Historical Society, Reno; Harold Hopkinson, Allis-Chalmers Corp., Milwaukee, Wis.; Donald Howe, Director of Public Relations, James B. Dunn, Asst. Dir. of Public Affairs, Homestake Mining Co., Lead, S.D.; Mrs. Marion Holmes, San Mateo County Historical Association, San Mateo, Calif.; Everett Jackson, Medical Sciences, Smithsonian Institution, Washington, D.C.; Al Johnson, Seattle, Wash.; Nancy Knepper, Colorado School of Mines Museum, Golden; Christine Lacy, Curator, The Molly Brown House, Denver, Colo; Harriett Meloy, Librarian, Lory Morrow, Photo Archivist, Montana Historical Society, Helena; Marjorie Morey, Amon Carter Museum, Fort Worth, Texas; E. W. Nolan, Librarian, Seattle Historical Society, Seattle, Wash.; Arnold Olean, Gold Mine Silver Screen, Black Hawk, Colo.; Arthur Olivas, Photographic Archivist, Richard Rudisill, Museum of New Mexico, Santa Fe; Margery Pontius, Librarian, Deadwood Public Library, Deadwood, S.D.; George Roe, Silver Plume, Colo.; LaVerne B. Rollin, Technical Editor, Mackay School of Mines, University of Nevada, Reno; Katherine Thornby, Curator, Adams Memorial Museum, Deadwood, S.D.; Dr. Lesta Turchen, Department of History, Dakota Wesleyan University, Mitchell, S.D.; Nelson Wadsworth, Salt Lake City, Utah; Elvira Wunderlich, Telluride, Colo.

BIBLIOGRAPHY

Addenbrooke, Alice B., *The Mistress of the Mansion*. Reno, 1950.

Adney, Tappan, *The Klondike Stampede*. Harper & Bros., 1899.

Andrews, Ralph W., *Picture Gallery Pioneers*. Bonanza Books, 1964.

Andrist, Ralph, *The Long Death*. The Macmillan Company, 1974.

Bancroft, Caroline:
  Silver Queen: The Fabulous Story of Baby Doe. The Golden Press, Denver, 1952.
  The Unsinkable Mrs. Brown. Johnson Publ. Co., Boulder, 1963.
  "Cousin Jack Stories from Central City," *Colorado Magazine*, Vol. 21, No. 2, March 1944.

Becker, Ethel Anderson, *Klondike '98*. Binfords & Mort, 1967.

Beebe, Lucius and Charles Clegg, *Legends of the Comstock Lode*. Graham H. Hardy, Oakland, Calif., 1950.

Berton, Pierre, *The Klondike Fever*. Alfred A. Knopf, 1958.

Billington, Ray Allen:
  The Far Western Frontier. Harper & Row, 1956.
  Westward Expansion. The Macmillan Company, 1967.

Blair, Edward, *Palace of Ice: A History of Leadville's Ice Palace 1895-1896*. Timberline Books, Leadville, Colo., 1972.

Blair, Edward, and E. Richard Churchill, *Everybody Came to Leadville*. Timberline Books, Leadville, Colo., 1971.

Blasters Handbook: A manual describing explosives and practical methods of use. Explosives Department, E. I. duPont de Nemours, 1966.

Blower, James, *Gold Rush*. American Heritage Press, 1971.

Brown, Dee, *The Gentle Tamers*. G. P. Putnam's Sons, 1958.

Bueler, Gladys R., *Colorado's Colorful Characters*. The Smoking Stack Press, Golden, Colo., 1975.

Burgess, Hubert, "Anecdotes of the Mines," *Century Magazine*, Vol. 42, 1891.

Burke, John, *The Legend of Baby Doe*. G. P. Putnam's Sons, 1974.

Chafetz, Henry, *Play the Devil*. Clarkson N. Potter, Inc., 1960.

Church, John A., "Accidents in the Comstock and Their Relation to Deep Mining," *Transactions of the American Institute of Mining Engineers*, Vol. 8 (1879).

Clark, Walter Van Tilberg, ed., *The Journals of Alfred Doten, 1849-1903*. 3 Vols. University of Nevada Press, 1973.

Clifford, Howard, *The Skagway Story*. Alaska Northwest Publishing Co., 1975.

Colby, Merle, *A Guide to Alaska*. The Macmillan Company, 1940.

Crampton, Frank A., *Deep Enough*. Sage Books, 1956.

Dallas, Sandra, *Cherry Creek Gothic: Victorian Architecture in Denver*. University of Oklahoma Press, 1971.

Dawson City News, December 15, 1897.

De Quille, Dan, *The Big Bonanza*. Alfred A. Knopf, 1947.

Dorset, Phyllis Flanders, *The New Eldorado*. The Macmillan Company, 1970.

Drago, Harry S., *Notorious Ladies of the Frontier*. Dodd, Mead, 1969.

Drury, Wells, *An Editor on the Comstock Lode*. Farrar & Rinehart, Inc., 1936.

En Route to the Klondike. W. B. Conkey Company, 1897.

Feitz, Leland, *Cripple Creek!* Little London Press, 1967.

Fielder, Mildred:
  The Treasure of Homestake Gold. North Plains Press, Aberdeen, S.D., 1970.
  Wild Bill and Deadwood. Bonanza Books, 1965.

Fisher, Vardis, and Opal Laurel Holmes, *Gold Rushes and Camps of the Early American West*. Caxton Printers, 1968.

Glasscock, C. B.:
  The Big Bonanza. Bobbs-Merrill, 1931.
  Gold in Them Hills. Bobbs-Merrill, 1932.

Greever, William S., *The Bonanza West*. University of Oklahoma Press, 1963.

Griswold, Don and Jean, *The Carbonate Camp Called Leadville*. University of Denver Press, 1951.

Hafen, LeRoy R., ed., *Pike's Peak Gold Rush Guidebooks of 1859*. The Arthur H. Clark Company, 1941.

Harvey, Mrs. James R., "The Leadville Ice Palace of 1896," *Colorado Magazine*, Vol. 17, May 1940.

History of Idaho Territory. Wallace W. Elliott & Co., 1884.

Horan, James D., *Timothy O'Sullivan, America's Forgotten Photographer*. Doubleday & Co., 1966.

Hughes, Richard B., *Pioneer Years in the Black Hills*. The Arthur H. Clark Company, 1957.

Hunt, William K., *North of 53°*. The Macmillan Company, 1974.

Jackson, Donald, *Custer's Gold*. University of Nebraska Press, 1972.

Jim Wardner of Wardner, Idaho by Himself. Facsimile Reproduction. The Shorey Books Store, Seattle, Washington, 1971.

Johns, William Douglas, *The Early Yukon, Alaska & the Klondike Discovery*. Mss. in the Univ. of Washington library, dated 1941.

Kohl, Edith Eudora, *Denver's Historic Mansions*. Sage Books, 1957.

Krause, Herbert and Gary D. Olson, *Custer's Prelude to Glory*. Brevet Press, 1974.

Ladue, Joseph, *Klondyke Facts*. American Technical Book Co., 1897.

Leighton, Caroline C., *Life at Puget Sound*. Lee and Shepard, 1884.

Leonard, John W., *The Gold Fields of the Klondike*. A. N. Marquis and Co., 1897.

Lewis, Oscar, *Silver Kings*. Alfred A. Knopf, 1967.

Lewis, Sidney, *The Labor Wars, From the Molly Maguires to the Sitdowns*. Doubleday & Co., 1973.

Lingenfelter, Richard E., *The Hardrock Miners*. University of California Press, 1974.

Lord, Eliot, *Comstock Mining and Miners*. Howell-North, 1959. (Reprint of 1883 edition.)

Lyman, George D., *The Saga of the Comstock Lode*. Charles Scribner's Sons, 1934.

MacDonald, Alexander, *In Search of El Dorado*. T. Fisher Unwin, 1905.

"A Manzama Heads North: Letters of William A. Langille." Lawrence Rakestraw, ed., *Oregon Historical Quarterly*, Vol. 76, No. 2, June 1975.

Marcosson, Isaac F., *Anaconda*. Dodd, Mead, 1957.

Martin, Cy, *Whiskey and Wild Women*. Hart Publishing Co., 1974.

Martinsen, Ella Lung, *Black Sand & Gold*. Binfords & Mort, 1967.

Mathews, Richard, *The Yukon*. Holt, Rinehart & Winston, 1968.

McLean, Evalyn Walsh, with Boyden Sparkes, *Father Struck It Rich*. Little, Brown, 1936.

Morgan, Murray, *One Man's Gold Rush*. University of Washington Press, 1967.

Morrell, W. P., *The Gold Rushes*. The Macmillan Company, 1941.

Mumey, Nolie, *Clark Gruber and Co. 1860-65*. Artcraft Press, 1950.

Naef, Weston, and James N. Wood, *Era of Exploration*. New York Graphic Society, 1975.

O'Connor, Richard, *High Jinks on the Klondike*. Bobbs-Merrill, 1954.

"Overland to Pike's Peak with a Quartz Mill, Letters of Samuel Mallory." *Colorado Magazine*, Vol. 8, 1931.

Paine, Swift, *Eilley Orrum, Queen of the Comstock*. Bobbs-Merrill, 1929.

Parker, Watson, *Gold in the Black Hills*. University of Oklahoma Press, 1966.

Paul, Rodman Wilson:

California Gold. University of Nebraska Press, 1947.
Mining Frontiers of the Far West 1848-1880. University of New Mexico Press, 1974.

Phillips, James W., *Alaska-Yukon Place Names*. University of Washington Press, 1963.

Quiett, Glenn Chesney, *Pay Dirt*. D. Appleton-Century, 1936.

Ray, Grace E., *Wily Women of the West*. The Naylor Co., 1972.

Rickard, T. A.:

A History of American Mining. McGraw-Hill, 1932.
"Rich Ore and its Moral Effects," *Mining and Scientific Press*, June 6, 1908.
"Salting," *Engineering and Mining Journal*, March 1941.

Ridge, Martin, and Ray Allen Billington, *America's Frontier Story*. Holt, Rinehart & Winston, 1969.

Roberts, R. W., *A Tramp to the Klondike; Or How I Reached the Gold Fields of Alaska*. Vaughnsville, Ohio.

Rowse, A. L., *The Cousin Jacks*. Charles Scribner's Sons, 1969.

Sage, Walter N., ed., "Record of a Trip to Dawson" (1898 diary of John Smith). *British Columbia Historical Quarterly*, Vol. 16, 1952.

Satterfield, Archie, *Chilkoot Pass*. Alaska Northwest Publ. Co., 1973.

Saxton, Alexander, *The Indispensable Enemy: Labor and the Anti-Chinese Movement in California*. University of California Press, 1971.

Schell, Herbert S., *History of South Dakota*. University of Nebraska Press, 1961.

Shinn, Charles Howard, *Mining Camps*. Alfred A. Knopf, 1948.

Sloane, Howard N. and Lucille L., *A Pictorial History of American Mining*. Crown Publishers, Inc., 1970.

Smith, Duane A.:

Horace Tabor: His Life and the Legend. Colorado Associated University Press, Boulder, 1973.
Rocky Mountain Mining Camps. Univ. of Nebraska Press, 1967.

Smith, Grant H., *The History of the Comstock Lode 1850-1920*. Nevada Bureau of Mines & University of Nevada, 1943.

Spence, Clark C., *Mining Engineers & The American West*. Yale University Press, 1970.

Sprague, Marshall, *Money Mountain*. Little, Brown, 1953.

Stanton, James B., *Ho for the Klondike*. Hancock House, 1974.

Stewart, Edgar L., *Custer's Luck*. University of Oklahoma Press, 1971.

Twain, Mark, *Roughing It*. New American Library, 1962.

Van Gelder, Arthur Pine, *History of the Explosives Industry in America*. Columbia University Press, 1927.

Villard, Henry, *The Past and Present of the Pike's Peak Gold Regions*. Da Capo Press, 1972.

Walker, Franklin, *Jack London of the Klondike*. The Huntington Library, 1972.

Watkins, T. H., *Gold and Silver in the West*. American West Publishing Co., 1971.

Wharton, David, *The Alaska Gold Rush*. Indiana Univ. Press, 1972.

Will C. Barnes' Arizona Place Names, revised by Byrd H. Granger. The University of Arizona Press, 1975.

Willison, George F., *Here They Dug the Gold*. Reynal & Hitchcock, 1946.

Winslow, Kathryn, *Big Pan-Out*. W. W. Norton & Co., 1951.

Wolle, Muriel Sibell, *The Bonanza Trail*. Indiana Univ. Press, 1953.

Young, Otis E., Jr.:

Western Mining. University of Oklahoma Press, 1970.
Black Powder and Hand Steel, Mines and Machines on the Old Western Frontier. University of Oklahoma Press, 1975.